CW00836448

Researching in the Former Soviet Union

Written for early-career scholars still in the planning stages of their research, this book explores some of the challenges researchers face when conducting fieldwork in the former Soviet region. It addresses key questions, including: What difficulties do scholars, especially females, encounter when researching in the region? How does an early-career scholars' positionality – especially their nationality, ethnicity, and sexuality – contribute to their experiences of inclusion, exclusion, and access while conducting fieldwork? How do early-career scholars navigate issues of personal safety in the field? How do junior academics successfully conduct high-risk research? The book includes contributors from both the region and Western countries, paying particular attention to the ways researchers' subjectivities shape how they are received in the region, which, in turn, influence how they write about and disseminate their research. The book also explores ways to continue research away from the field through the use of digital methods when physical access is not possible.

Jasmin Dall'Agnola is an Associated Research Fellow at the Organization for Security and Cooperation in Europe (OSCE) Academy in Bishkek. Her research centers on the relationship between gender, governance, and technology in post-Soviet Central Asia. She has been awarded a prestigious two-year, full-time Postdoc. Mobility Fellowship from the Swiss National Science Foundation, to explore the implications of the COVID-19 pandemic on smart city technologies in Central Asia. In her role as founder and acting chair of the Eurasian, East and Central European Studies Women Academics Forum (EECES WAF), Jasmin is involved in various collaborative research and networking endeavors to address gender disparity in academia.

Allyson Edwards research examines Militarism in Post-Soviet Russia, more specifically, the mechanisms behind latent militarization between 1990 and 2000. She is currently a lecturer in Global History at Bath Spa University and plays a wider role in the academic community as Vice-Chair of the EECES WAF. She specializes in topics related to militarism, memory, education, and parades in the Russian/Eurasian space.

Marnie Howlett is a Departmental Lecturer in Politics in the Department of Politics and International Relations at the University of Oxford. She has held Fellowships supported by the Social Sciences and Humanities Research Council of Canada (SSHRC), the Canadian Foundation for Ukrainian Studies, and the John Fell Fund. Marnie was a Deputy Editor for *Millennium: Journal of International Studies*, vol. 48 and currently sits on the editorial board for *Qualitative Research*. Her research centers on the intersection of nationalism, geopolitics, and cartography within the former Soviet Union, particularly Ukraine.

BASEES/Routledge Series on Russian and East European Studies

Series editors:

This series is published on behalf of BASEES (the British Association for Slavonic and East European Studies). The series comprises original, high-quality, research-level work by both new and established scholars on all aspects of Russian, Soviet, post-Soviet and East European Studies in humanities and social science subjects.

147 Projecting Russia in a Mediatized World
Recursive Nationhood
Stephen Hutchings

148 Russia's Regional Museums
Representing and Misrepresenting Knowledge about Nature, History and Society
Sofia Gavrilova

149 Russian Nationalism
Imaginaries, Doctrines, and Political Battlefields
Marlene Laruelle

150 Researching in the Former Soviet Union
Stories from the Field
Edited by Jasmin Dall'Agnola, Allyson Edwards and Marnie Howlett

For a full list of available titles please visit: https://www.routledge.com/BASEES-Routledge-Series-on-Russian-and-East-European-Studies/book-series/BASEES

Researching in the Former Soviet Union

Stories from the Field

Edited by
Jasmin Dall'Agnola, Allyson Edwards, and Marnie Howlett

LONDON AND NEW YORK

First published 2023
by Routledge
4 Park Square, Milton Park, Abingdon, Oxon OX14 4RN

and by Routledge
605 Third Avenue, New York, NY 10158

Routledge is an imprint of the Taylor & Francis Group, an informa business

British Library Cataloguing-in-Publication Data
A catalogue record for this book is available from the British Library

ISBN: 978-0-367-69993-2 (hbk)
ISBN: 978-0-367-69995-6 (pbk)
ISBN: 978-1-003-14416-8 (ebk)

DOI: 10.4324/9781003144168

Typeset in Times New Roman
by KnowledgeWorks Global Ltd.

Contents

Contributors

Rasa Kamarauskaitė is a sociology PhD student at School of Slavonic and East European Studies (UCL). Her research focuses on everyday (in)visibilities of non-heterosexual people in Lithuania. By the means of fieldwork, she seeks to examine the circumstances, conditions, and practices that render Lithuanian non-heterosexual people (in)visible in their everyday contexts. Thematically and conceptually her research is informed by studies of public and private, body and space, sexuality, communism and post-communism. In terms of methodology, she adheres to the principles of grounded theory.

Abigail Karas recently completed her DPhil at the University of Oxford. Her research looks at urban space, local identity, and municipal politics. Her doctoral thesis focused on the different uses of the city's rooftop (and attic) spaces and related these to social development, urban planning, civic activism, and the local imaginary in St Petersburg.

Andrea Peinhopf holds a PhD in Sociology and Anthropology from the School of Slavonic and East European Studies, University College London. Her PhD thesis, which was funded by the UK's Economic and Social Research Council (ESRC), explored the impact of violence and unresolved conflict on social relations in the contested state of Abkhazia, where she conducted long-term ethnographic fieldwork. The results of her research have been published in *Ethnopolitics*, *Nationalities Papers*, *The Palgrave Encyclopedia of Peace* and *Conflict Studies* and *The Global Encyclopedia of Informality*. In 2018, Andrea was a recipient of the Best Doctoral Paper Award by the Association for the Study of Nationalities (ASN). She is currently an ESRC Postdoctoral Research Fellow in the Department of Politics at the University of York, where she works on a project exploring the social impact of international isolation in the context of contested statehood.

Ruta Skriptaite is a doctoral researcher in the School of Politics and International Relations at the University of Nottingham. Her PhD research explores the affinity between masculinities and political image-making in the post-Soviet space, by providing a comparative analysis of the Belarusian president Aliaksandr Lukashenka, recently resigned Kazakh president Nursultan Nazarbayev, and deceased Turkmen president Saparmurat Niyazov-Turkmenbashi. Prior to starting her PhD research, Ruta completed an MA in Politics, Security and Integration at the School of Slavonic and East European Studies (UCL) and a BA (Hons) in International Relations at the University of Essex.

Hélène Thibault holds a PhD in Political Science from the University of Ottawa and is Assistant Professor at the faculty of Sciences and Humanities at Nazarbayev University since 2016. She specializes in ethnography, religion, secularism, and the Soviet legacy. Her current projects also look at gender issues, marriage, sexuality, and polygyny in Central Asia. She is also the co-director of the Central Asia hub of the Political Economy of Education Network (PEER), a transnational three-year project funded by the Global Challenges Research Fund (GCRF).

Zhaniya Turlubekova holds a dual doctorate degree from Utrecht University and Eötvös Loránd University (Summa Cum Laude) in Cultural and Global Criminology. Her PhD research was awarded a prestigious four-year full-time Erasmus Mundus Joint Doctorate Scholarship. She received a bachelor's degree in law from Kostanay State University and a master's degree in political science and international relations from Nazarbayev University. Apart from her academic experience, she has accomplished several internships and gained practical legal and analytical experience with a number of agencies at national and international levels.

Colleen Wood is a PhD student in the Department of Political Science at Columbia University. Her research on civic identity and political participation in Central Asia is supported by the National Science Foundation and the Harriman Institute. Prior to graduate studies, she earned a BS from Georgetown University and served as a Peace Corps volunteer for two years in Kyrgyzstan.

Preface

Hélène Thibault
Nazarbayev University

This edited volume sheds light on the realities and challenges of fieldwork for early-career female researchers working in the former USSR. This book will be useful for students and supervisors alike to think about some of the perils scholars can be exposed to when conducting fieldwork and help in developing methodological strategies in low- to high-risk environments. This preface is an opportunity for me to discuss the particularities of the former Soviet space as a research ground as well as the virtue and importance of reflexivity and positionality.

When it comes to research environments, every country, region, or community, has its own peculiarities and set of challenges. The former USSR is a very diverse territory, and some regions represent accessible, unregimented research spaces whereas others are characterized by authoritarian environments which limit informants' and researchers' freedom. It affects research even if it is conducted online as Ruta Skriptaite's experience reveals since some of her Turkmen informants cannot escape electronic surveillance. In this book, the cases of Lithuania and Abkhazia offer very contrasting pictures; the former is a peaceful country and member of the European Union, while the latter is an unrecognized *de facto* state that has been in limbo since the end of the civil war in Georgia in 1993 (see Andrea Peinhopf's story in this volume). Yet despite all their differences, countries of the region bear some similarities because of their shared historical path.

The former Soviet space is idiosyncratic because of the double colonization experience: Russian and Soviet. And the USSR was no ordinary empire. As a country, the Soviet Union was exceptional in many respects, because of its size, repressive politics, totalizing communist ideology, socioeconomic achievements, and influence on the world scene. Unlike some other colonial powers, the Soviet power had a double objective: economic extraction coupled with the complete transformation of the social and political order. Sociopolitical transformation under the Soviet regime is undeniable but the extent to which people were influenced by communist values or an authoritarian mindset is still subjected to intense debate. The legacy of Soviet rule is significant and continues to impact socioeconomic dynamics but this impact is felt differently depending on the region, in terms of political institutions, the strength of civil societies, or border issues. The Baltic states, for instance, have moved further away from their Soviet past than any other former republics. The fact that they had stronger political institutions prior to joining the USSR, and their proximity to Europe certainly contributed to their greater independence. Their open political systems and liberal societies facilitate access to the field. In contrast, Central Asia, parts of Russia, and the Caucasus appear as more challenging research fields because of political instability or authoritarian surveillance. In those contexts, informants might be more reluctant to engage with researchers, afraid to compromise their safety.

In terms of research, one reason why the former Soviet space stands out is that apart from few exceptions, the countries that made the Soviet Union were, for a long time, mostly off-limits to foreign scholars whereas ideological orientations and authoritarian politics were limiting the type of research local scholars could undertake. As a result, knowledge about the actual (as opposed to idealized) sociopolitical dynamics of this vast region was limited and little was known about its regional particularities and the hundreds of different ethnic and religious communities. Even today, the general Western public often talks about "Russia" when in fact referring to the Soviet Union. As the USSR fell apart in 1991 and countries became independent, both local and foreign scholars started investigating and disaggregating the profound changes that local communities were going through.

Another particularity of the region is connected to the Soviet women's emancipation project that continues to impact the gender order in the region (for more, see Peshkova and Thibault, 2022). The Soviet state's gender policies were strongly heteronormative and this legacy lives on in various degrees. Women's literacy rates and participation in the workforce are very high considering the level of economic development in some of the former USSR countries. Even though women tend to be underrepresented in decision-making circles, they are socially and publicly active and it makes it easier for female scholars to go around doing research even if they are sometimes perceived as inexperienced and patronized by older people and some men. Abigail Karas (in this volume), when conducting her doctoral research in Russia, found that being a foreign female was an advantage. She experienced that, in general, as a woman she was seen as less threatening to anyone, inspired more confidence and enjoyed easier access to respondents, both male and female. My field research experience matches her observation. During my own fieldwork with religious conservative communities in Khujand, Tajikistan, being a foreign unmarried non-Muslim female was an advantage. My informants used my status as a single foreign woman to steer interactions toward those of my religious conversion and need for marriage. Their repeated efforts and our interactions exposed the depth of their religious beliefs and its precedence over other identity markers such as ethnicity and language (for more, see Thibault 2021). Even if we are conscious of our own positionality, it is hard to fully grasp its impact on the ground. For instance, I can only guess on how differently my fieldwork would have unfolded if I had been a married practicing Muslim female scholar who belonged to the same community. How differently would people have reacted to me and to my inquiries then? We can only speculate but what matters is to be conscious that the researcher is biased, that interviewees are partial and that our position can represent an advantage or a disadvantage depending on circumstances.

Engaging in research that involves other humans is always a difficult and subtle endeavor that requires researchers to be brave and reach out to strangers and sometimes face rejection. It also requires great listening skills (see Colleen Wood's story in this volume) and tact in order to gain someone's trust and convince informants, people from whom we seek knowledge, to share their life stories or lifestyles with us for our *own* purpose. We require from our informants to sometimes share very intimate details of their lives which can cause discomfort. In this volume, some of the topics addressed are tough ones: war memories, sexual identity, nationalism, and drug trafficking, and we might assume that people will be reluctant to share with some of their opinions or life stories with strangers. But researchers themselves are sometimes surprised as to how easy it becomes to get "people to talk." It turns out that many humans like to talk about themselves and the realities they face. They do so for different reasons. They might have a political agenda and a cause to put forward. They might feel that researchers need to be educated on certain topics. Other times, it can be therapeutic for informants to talk to someone who has a genuine interest in them and their activities/life stories. It is also often the case that informants see the researcher as a neutral confidant who will not spill secrets. The construction of knowledge goes both ways, as the contributions in this collection demonstrate. It is as much as who we are than who we talk to.

Scholars tend to study people and groups who are less privileged than themselves. This is a rather uncomfortable truth because most academics do not feel that they take advantage of the informants from whom they seek knowledge. Indeed, it is not about taking advantage, it is about access. When scholars from the Global North conduct research in foreign countries that have a lower level of economic development, they are simply more privileged than most citizens, even when they are still students; they are mobile, they have visas, they have access to funding and are often less subjected to social norms and, to some extent, less likely to get arrested because their detention could attract unwanted international attention. When conducting research in Tajikistan and Kazakhstan, some of my informants claimed to be more disposed to talk to me *because* I was a foreigner. Their rationale was that I had limited acquaintances in the community and would therefore be less likely to gossip about them.

When researchers are *from there*, however, the dynamics are a little different. There are very few accounts of researcher reflecting on their research experience in a state they have family ties to. In their collection of articles published in *openDemocracy* in 2019, female scholars from Central Asia discuss some of the challenges and dangers they face when doing research *at home* (Kudaibergenova 2019). In it, Mohira Suyarkulova (2019) argued that local researchers can be met with incredulity by their fellow citizens who are wondering why the local researcher does not already know the answers to all the questions she asks to potential informants. Even though some of the scholars in this volume are natives to the region, most of them were affiliated with Western institutions during their doctoral studies. They are locals, but a different kind of local as Marnie Howlett remarked. Being a different kind of local, in Howlett's case a member of the Ukrainian diaspora in Canada, turned out to be very beneficial for her data collection in Ukraine.

Many of the countries of the former Soviet Union are characterized by authoritarianism and some of the research scholars engage in can become problematic for them and/or their informants. However, when you are *from* there and do not have a status in a different country (Howlett has a Canadian passport), you cannot leave if you get into trouble, you must face the consequences of your involvement with local dynamics. Even worse, perhaps family or friends will also suffer the consequences of research that disturbs local politics. The case of Alexander Sodiqov, a Tajikistani citizen studying in Canada who was arrested on charges of espionage when conducting research in Tajikistan on behalf of an international team in 2014, is one of the most dramatic examples of the perils faced by local scholars, although international pressure and diplomatic efforts secured his release few months later (Clibbon 2014). These are challenges Zhaniya Turlubekova (in this volume), as a Central Asian female scholar, can certainly relate to since her research on drug trafficking in her home country, Kazakhstan, concerns a very sensitive topic and she was sometimes perceived as a "spy" by some informants. Rasa Kamarauskaite, likewise, was vulnerable as a lesbian Lithuanian researcher focusing on nonheterosexual communities in Lithuania, a country that is known to be one of the most homophobic countries in the European Union. Despite the dangers, it seems that they managed to adroitly navigate their fieldwork without compromising their safety. As more and more scholars around the world initiate discussions about racial inequality and hierarchies of knowledge, reflecting on positionality seems even more important.

Positionality and reflexivity have had an increased presence in social science research in the last few decades. Most universities' ethics boards ask researchers to pay a great deal of attention to the impact that their research could potentially have on participants. This process certainly helps scholars envision their fieldwork more clearly and reflect on the possible impacts, negative or positive, of their presence on local realties. Yet, these ethics requirements sometimes appear to be meant to protect universities from legal pursuits more than to provide guidance for research. These ethics requirements often do not come with a lot of guidance for researchers. If graduate students are lucky, they will have access to qualitative methodology courses that address the challenges of fieldwork, but these are relatively rare, especially in North American universities. If they are luckier, someone who has extended experience in doing field research will supervise them, but this is not always the case. Managing fieldwork is a very delicate enterprise and universities could definitely put more efforts into preparing their students for field research and highlighting the challenges they will face. Most importantly, they should recognize the different positions young researchers find themselves in depending on their class, ethnic, or sexual identity. This book, with its recollection of research experiences that emphasize positionality and intersectionality, contributes to demystifying fieldwork and provides much-needed guidance to early-career female researchers who are preparing to conduct field research.

While most academic ethics committees require scholars to outline the impact they have on the dynamics and lives of the people from whom they seek knowledge, this aspect is seldom exposed in academic literature. In traditional, positivist social science, researchers are supposed to be neutral and aim to have as little impact as possible on the "reality" that is being studied. References to researchers' own values and positions were frowned upon and considered unprofessional. The pushback that this edited volume received by reviewers is a testimony of that resistance to incorporate personal narratives into academic writing. Criticized for their honesty, perceived lack of ethics and recklessness, the contributors have on the contrary offered us a very sincere account of what research looks like in difficult environments and honest reflections on gendered positionality. Being honest and transparent about our interactions in the field does not mean the research is unscientific or unreliable. On the contrary, it allows researchers to detail certain hidden aspects of the fieldwork and how it can affect data collection and the scientific reflection, which can only give credibility to the research process.

References

Clibbon, J. 2014. "How Alexander Sodiqov Was Freed Following Espionage Charges." *CBC*, September 13. https://www.cbc.ca/news/world/how-alexander-sodiqov-was-freed-following-espionage-charges-1.2772191.

Kudaibergenova, D. 2019. "When Your Field Is Also Your Home: Introducing Feminist Subjectivities in Central Asia." *openDemocracy*, October 7. https://www.opendemocracy.net/en/odr/when-your-field-also-your-home-introducing-feminist-subjectivities-central-asia/.

Peshkova, S., & Thibault, H. 2022. "Introduction." *Central Asian Affairs* 9 (2–4): 149–175. DOI: 10.30965/22142290-12340019.

Suyarkulova, M. 2019. "A View from the Margins: Alienation and Accountability in Central Asian Studies." *openDemocracy*, October 10. https://www.opendemocracy.net/en/odr/view-margins-alienation-and-accountability-central-asian-studies/.

Thibault, H. 2021. "'Are You Married?': Gender and Faith in Political Ethnographic Research." *Journal of Contemporary Ethnography* 50 (3): 395–416. DOI: 10.1177/0891241620986852.

A Note from the Editors

The idea for this volume emerged following a roundtable discussion about emerging female scholars' fieldwork experiences researching post-Communist states, held in London in February 2020. Clearly, nationality and other identifying characteristics play a key role in the way researchers are received in the region, especially early-career scholars, which, in turn, influences how they write about and disseminate their research. The book offers useful reflections and insights for others researching the region, especially those still at the planning stage for their field research. The editors hope that the insights offered in this volume will be a valuable source for those teaching qualitative research methodology, with a focus on courses covering fieldwork preparation and planning within the social sciences.

Introduction

The Challenges of Fieldwork in Post-Soviet Societies

Jasmin Dall'Agnola
OSCE Academy

The principal aim of this volume is to start an honest discussion about the practical, methodological, and ethical challenges early-career female scholars face while conducting field research in the states of the former Soviet Union. So far, only a handful of scholars have written openly about the unique obstacles that the region presents to early-career researchers, especially young women, obstacles which are often not encountered in other field sites (Adams 1999; Allina-Pisano 2009; Johnson 2009; Thibault 2021).

It is this gap in the literature that the present collection seeks to address. In this volume, young early-career female fieldworkers speak about their experiences conducting PhD fieldwork[1] in the post-Soviet region. All the authors have recently, or will soon, complete their doctoral studies. While documenting some of the challenges they faced during their PhD field research, the contributors engage with their positionality and intersectional identities in the field.[2] The chapters therefore present honest and reflective accounts of the authors' personal experience in the field. The contributors adopted various methodologies for their PhD projects (e.g., oral history interviews, participant observations, focus groups, digital ethnography), as well as a variety of topics, including drug trafficking, activism, roofing, nationalism, and gender. Moreover, the scope of this volume is not only situated in the range of geographical countries under analysis, but also the authors represent diverse disciplines in the social sciences (e.g., anthropology, ethnography, sociology, political science, criminology) and nationalities and are affiliated with numerous academic institutions located in different parts of the world.

The chapters present personal stories – quite deliberately. There is no shortage of comprehensive guides on field research methodology and all the contributors to this volume have made use of those. The insights gained from these textbooks and various pre-fieldwork workshops have helped them to develop a coherent methodological approach to addressing their research questions, as well as to complete detailed risk assessment and ethical clearance forms prior to their field trips. This volume demonstrates that, despite thorough preparations, scholars may encounter situations in the field (e.g., political or economic turmoil, health emergencies, and environmental disasters) requiring them to adjust their research methodology

DOI: 10.4324/9781003144168-1

and, in some cases, even their research topics. The COVID-19 pandemic in particular has significantly disrupted research globally. Yet, it has also shown that scholars do not have to leave their homes, countries, or institutions in order to conduct fieldwork (see, for example, Howlett 2021). Still, the practical and ethical dilemmas faced by scholars who are conducting remote fieldwork,[3] especially within the former Soviet states, have yet to be comprehensively addressed. Through the medium of firsthand accounts of PhD fieldwork, this collection does just that. It discusses the discrepancies between textbook advice and the practical conduct of on- and offline field research observed by early-career female scholars.

The volume is organized chronologically with regard to the influence of the COVID-19 pandemic on contributors' fieldwork in the region. The book consists of three parts. Part I (Stories from the Post-Soviet Field) presents personal accounts from women researchers who finished their PhD field research in the post-Soviet region prior to the start of the COVID-19 pandemic. It illustrates some of the challenges that early-career female scholars faced when conducting their PhD fieldwork in the region in pre-COVID times. In opening the book with a discussion of her field research experiences in Abkhazia, Andrea Peinhopf calls on scholars not to lose heart when they meet certain challenges during data collection in the field. While reflecting on her own experiences of betrayal (Chapter 1), she offers comfort to other female researchers who plan to conduct research in *de facto* states of the former USSR, highlighting the invaluable insights that perceived "failures" can generate. Rasa Kamarauskaitė pushes the discussion further in the following chapter (Chapter 2) by questioning what happens when a scholar's "home" is in fact her field site. She does so by showing how her Lithuanian and queer identity helped her to bond with nonheterosexual respondents during her PhD field research in Lithuania. Likewise, Zhaniya Turlubekova, as a Kazakh woman, was somehow returning "home" when embarking on her PhD research on marijuana and heroin trafficking in Kazakhstan. In Chapter 3, she refashions our understanding of the practical and ethical dilemmas early-career female scholars face while conducting high-risk research "at home," in an authoritarian post-Soviet state.

While Peinhopf, Kamarauskaitė, and Turlubekova finished their PhD fieldwork prior to the outbreak of the coronavirus in December 2019, the second part of this volume (Stories from the Hybrid Field) consists of two female fieldworkers who have conducted both off- and online PhD field research in the region as a consequence of the COVID-19 pandemic. Both stories demonstrate that in prompting early-career researchers to move their field research online COVID-19 has helped some of them to diversify their methodological approach and in doing so has enriched the data collection for their doctoral studies. While in pre-COVID times, Karas' respondents only reluctantly engaged in virtual communication with her, the fact that they were isolated and bored during the first global lockdown seemed to make them more at ease and willing to talk online. In Chapter 4, she shows how being a white British female researcher meant that, in many ways, she was uniquely well positioned to study the urban exploration community in St. Petersburg in Russia, both on- and offline. Similar observations have been reported by Marnie Howlett (Chapter 5) in this collection. In her chapter, Howlett speaks about the significance of her identity in gaining access to the field by demonstrating how her Ukrainian-Canadian diaspora connection proved helpful in gaining her Ukrainian interlocutors' trust, even when COVID-19 forced her to switch to remote fieldwork. Despite sharing ancestry and an understanding of cultural nuances, Howlett was still seen by her participants as an "inbetweener" (Milligan 2016, 248), i.e., neither a true insider nor an outsider.

The third part of this volume (Stories from the Digital Field) homes in on the strategies employed by two early-career female researchers who had to rethink their field research methodology and were forced to move their PhD research fully online due to the pandemic. Both coped surprisingly well with the unknown situation. In Chapter 6, Wood describes how she had to readjust her initial field strategy when the Kyrgyz government closed the state's borders to foreigners in spring 2020. For her, digital ethnography did not pose restrictions per se but presented new, useful, and insightful angles to study civic activism in Kyrgyzstan. Her personal story offers useful insights to other researchers who are forced to study activism in the post-Soviet space through a digital lens. For Ruta Skriptaite (Chapter 7), likewise, online interviews with experts from Belarus, Kazakhstan, and Turkmenistan on their perception of masculinity and leadership turned out to be less time-consuming in terms of travel and presented fewer financial and social hurdles than in-person interviews. Her personal story in this collection illustrates some of the challenges both early-career and more senior female scholars with a working-class background face when trying to balance their job and remote fieldwork responsibilities with childcare obligations.

In the remainder of this introductory chapter, I present the three key themes that emerged from the various personal stories in this volume.

Self-Representation and Its Risks in Post-Soviet Societies

As the contributions to this volume show, managing self-representations when conducting research in the states of the former Soviet Union can be a particular challenge for scholars, and especially young women, as sexism is part of the nation-building discourse (see, for example, Thibault's preface in this volume). Women scholars stand out in these male-dominated societies. They often receive a lot of unwanted attention from, say, taxi drivers, shop owners, and receptionists, with unexpected inquiries and comments about why they are unmarried and without children (Thibault 2021).

Previous female scholarship on the former Soviet Union has carefully collected a range of safety precautions and advice to assist young women scholars conducting research in patriarchal societies (Adams 1999; Johnson

2009; Thibault 2021). For instance, to maximize their safety, female fieldworkers are often advised to wear androgynous, culturally appropriate (e.g., modest) clothing, to adopt local mannerisms, to wear a fake wedding ring, to use public transport rather than a private car or taxi, to restrict their fieldwork to the daylight hours, and to rely on local gatekeepers for their recruitment of research participants. While this well-meant guidance by established female scholars is important and insightful, the personal stories in this collection demonstrate that some of these suggestions need to be updated as they no longer speak to the contemporary reality on the ground. Some accounts presented here emphasize discrepancies, outdatedness, or even biases implicit in the advice often given about fieldwork to more junior scholars. For instance, when reflecting on her travel and work experiences in Kyrgyzstan, Wood discusses the US Peace Corps' safety training, which she felt served to reinforce some Western women's stereotypes about Central Asian men's assumed predatory behavior. Karas, likewise, found that some of the advice she had received from more senior researchers prior to entering the field turned out to be misleading and outdated, since her roofers, apart from some flirty jokes and remarks, treated her respectfully. Both Wood and Karas push back on a dominant scholarly account that focuses on the discourse of danger and difficulties faced by female researchers conducting fieldwork in male-dominated environments.

Moreover, the stories in this edited volume show that even when female field researchers decide to dress modestly, unwanted romantic and sexual advances by men cannot be precluded. Although, Howlett, Karas, and Wood adopted modest clothing and local language patterns to draw less attention to their foreign femininity, they were all confronted to some degree with flirtatious behavior in the form of jokes, gifts, and dinner invitations from their male respondents. As someone who has conducted her own PhD field research in the heteronormative society of Kazakhstan, I can certainly understand how challenging it must have been for them to navigate those situations. Because, despite the fact that I wore a fake engagement ring during my own fieldwork in Kazakhstan, men in the field regularly flirted with me, with one even asking me to become his second wife (*toqal*). In light of the persistent unwanted advances from Kazakh men, I decided to turn my attention to female informants and inquire about their experiences. Through these conversations with local women about their conception of marriage, family, and local gender norms, I soon learned that in the Kazakh tradition an engagement ring is worn on the right ring finger, while I was wearing the fake ring on my left ring finger, as is the custom in my home country (Switzerland). These conversations with female acquaintances helped me not only to navigate the romantic offers by Kazakh men in the field but also resulted in a focus on gender in my written work (Dall'Agnola 2020, 2021, 2022), even though my original PhD project had not included gender in any significant way. The exchanges also inspired me to initiate this edited volume on early-career female researchers' fieldwork experiences in post-Soviet states.

Furthermore, reliance on gatekeepers may carry its own (often gendered) risks, as Peinhopf's account shows. Although she dressed conservatively, the fact that she was a young single foreign "lone female researcher" (Johansson 2015, 59), without official affiliation to a local university, made her an easy target for betrayal. In contrast to most other contributors who focused on the researcher's position of power in the field, Peinhopf's story illustrates the influence that interlocutors (and especially local men) can have on female fieldworkers. Her account suggests that having an institutional affiliation – although not always feasible – can be helpful in mitigating some of the "gendered risks" (Johansson 2015, 55) that can arise from collaborations with people on the ground. Yet, in her chapter, Peinhopf also stresses that having an institutional affiliation might be problematic for other reasons, such as carrying the risk of reputational damage or facing pressures and compromising one's independence as a researcher. Moreover, her story illustrates another common quandary that women scholars can face when doing research in male-dominated societies; namely, how to ward off repeated romantic and sexual advances from informants and gatekeepers without offending them or losing their support (Mügge 2013; Johansson 2015). Turlubekova experienced a similar dilemma when one of her informants, a Russian Mafia-affiliated drug trafficker, offered her introductions to various wholesale heroin dealers in Kazakhstan if she agreed to continue their meeting in private. Overwhelmed by her respondent's demand, Turlubekova's initial reaction was to freeze. Her subsequent, carefully worded refusal was misread by her respondent as a classic trap that had been laid by law enforcement officials to obtain information from him and other interlocutors about past or planned crimes. As a result, the drug trafficker was no longer willing to offer his assistance.

Nevertheless, and in line with previous scholarship on field research, the contributions in this collection show that unmarried female scholars, and especially those who are young, not only stand out in male-dominated societies but also enjoy several advantages over male researchers (Tadevosyan 2019; Cromwell and Tadevosyan 2021; Thibault 2021). Importantly, it is not solely their "sexual availability" as unmarried women that can enhance the willingness of male respondents to engage with them. As the personal stories in this volume show, in some cases, being a woman can be particularly advantageous for data collection in male-dominated groups. Most notable is the fact that being a white British woman helped Karas to gain the trust of her roofers, who were predominantly male, in St. Petersburg; not only was her presence seen as nonthreatening, but her interlocutors' expectations toward her were also different and somewhat less demanding than toward men or other females in Russian society. Whereas a male researcher's declining of an offer of alcohol would have likely been met with skepticism and possible exclusion from future gatherings, Karas' objection to alcohol did not harm her social standing in the group. Peinhopf's experience of victimization, likewise, rather than harming her field research, offered useful insights into the local gender dynamics of her research participants in Abkhazia and enriched her data collection. The process of making sense of these insights in collaboration with locals in Abkhazia who have gone through similar experiences brought her closer to the mindset of an insider, thus softening the boundaries between insider and outsider status. Moreover, Peinhopf realized that her female gender helped mitigate the exposure to hostility, as it is considered less appropriate in Abkhazia to exhibit outright threatening or hostile attitudes toward a woman than a man.

Likewise, the contributions in this volume show that local scholars can also find their gender to be helpful in establishing a rapport with research participants. Skriptaite notes that one of her interactions with a female expert from Kazakhstan became more personal and insightful when it emerged that she, Skriptaite, was a mother. Nevertheless, it is unclear whether her respondent's changed attitude toward her was due to the glorification of motherhood in post-Soviet societies or to her own identity as a mother. As an expat Lithuanian living in London, Kamarauskaitė no longer identified herself as a local Lithuanian woman. Yet, despite her presumed outsider identity, her research participants clearly identified her as "one of them," namely, a fellow lesbian Lithuanian woman. Notably, being perceived as a socioeconomically vulnerable PhD student, in addition to her sexuality and her language capabilities, facilitated deep and rich reflections from her interviewees and marked Kamarauskaitė out as an insider. In a similar vein, Turlubekova argues that being a "beautiful Kazakh girl" facilitated her access to policemen, drug dealers, and drug lords because of the regional sexist stereotype that women and especially young women "are to be helped." Her data collection consequently benefited from this paternalistic treatment by her male respondents as they often unintentionally shared more insightful information on local drug trafficking practices with her than they would likely have done with male scholars.

Still, Turlubekova's experience differs from the two previously outlined, as her gender did not necessarily help her gain her female interlocutors' trust in Kazakhstan; instead, her privileged educational and social upbringing was a hindrance. Turlubekova's impression has been confirmed by other local female scholars in post-Soviet Central Asia (see, for example, Kudaibergenova 2019), who argue that it is not just gender but also the intersecting factors of class, culture, age, and race, which further complicate a researcher's positionality and thus interactions with interlocutors in the field. In summary, the contributors' personal stories highlight the dynamic and intersecting facets of early-career female scholars' identities in the field, and the ways they shape both the data collection and the outcome of the PhD research.

Physical Safety and Mental Well-Being

As is openly and honestly revealed in this collection through personal stories, research in the states of the former USSR can involve several unexpected dangers and risks for both research participants and fieldworkers, especially if that researcher is a woman. Although all authors had attended various forms of pre-fieldwork workshops and had completed detailed risk assessment and ethical clearance forms, they all experienced feelings of stress and insecurity while in the field due to the politically repressive nature of the region, their research topic, cultural nuances, isolation in being far away from friends and family, and the COVID-19 pandemic. In reflecting on their experiences in the field, they offer useful insights for other fieldworkers on how to maximize their respondents' as well as their own physical safety and mental well-being prior to, during, and after fieldwork. In addition, their personal stories highlight the need for doctoral students and their supervisors to run a careful risk-benefit analysis of the potential dangers that researchers may face when doing research in the post-Soviet region, especially when working in the field for the first time.

As the personal stories of Karas, Turlubekova, and Peinhopf in this collection show, scholars conducting high-risk research often face additional institutional ethics challenges that ask fieldworkers to carefully choose, and sometimes even to adapt, their research methods to secure the safety of both themselves and their respondents in the field. Studying activities defined as illegal or deviant in a scholar's home country raises a number of risks of physical harm to field researchers (Dekeyser and Garret 2021), as Turlubekova's story illustrates. To enhance her personal safety during and after her interviews with dealers and drug traffickers in Kazakhstan, Turlubekova set up several safety protocols with local friends and family members. For instance, she asked them to drive or pick her up from her meetings with high-profile criminals or to meet her in a café close to the interview location straight after the interview. Yet, as Karas' account demonstrates, the physical risks associated with high-risk research do not necessarily stem from research participants but can also derive from the studied activity itself. In her case, the activity studied, roofing, was an activity that is criminalized *de jure* in Russia. Because of the activity's illegality, and as engaging in it puts a researcher and her participants in great physical danger, Karas' University Research Ethics Committee ruled that it was too dangerous for her to climb onto the roofs herself. Not being able to participate in the activity of roofing herself, she searched for other forms of connection that would allow her to "hang out" with her participants by combining several qualitative methods: semi-structured interviews, participant observation, and digital ethnography. Similar physical risks can also apply to field researchers who conduct research in post-Soviet *de facto* states, for which their country's Foreign Office advises against all travel, as Peinhopf's personal account shows. Due to safety concerns, the British Foreign Office classified Abkhazia, Peinhopf's field site, as a no-go zone. To obtain approval, she therefore had to demonstrate to her university and ethics review board that she had considered – and was well-prepared for – all risks that could arise from her decision to travel to the *de facto* state. Her University Research Ethics Committee approved her field research provided she would continuously evaluate and assess the security situation along with international organizations and their local offices in Abkhazia and, if required, would be prepared to change her field strategy.

As fieldwork in the post-Soviet space may entail additional risks of harm for scholars who conduct research on and/or belong themselves to vulnerable or stigmatized groups (e.g., women, ethnic, religious, racial or sexual minority), risk assessments are especially important. For instance, for women of color, field research in some post-Soviet cities that are known for racial attacks can pose a greater risk to personal safety than for white female scholars (St. Julian-Varnon 2020). Likewise, scholars researching and belonging themselves to the lesbian, gay, bisexual, transgender, and queer (LGBTQ+) community, who work alone in the states of the former Soviet Union, where homosexuality continues to be either outlawed or violently suppressed by the state and society,[4] often have to take additional measures to protect their research participants and themselves from verbal and physical harm. These are sentiments Kamarauskaitė (in this volume), as a lesbian Lithuanian woman, can certainly relate to because Lithuanian society is known to be one of the most homophobic and transphobic in Europe (see, for example, ILGA-Europe 2022). Kamarauskaitė's story also demonstrates how important the anonymization of data and due care in handling it is, even more so when the researcher is studying a vulnerable and stigmatized group. As is outlined in her chapter, any involuntary "outing" of one of her respondents could have had negative repercussions for her interlocutors in Lithuania. Therefore, securing the anonymity and confidentiality of her participants' identities proved to be even more important for Kamarauskaitė, due to the potential harms associated with her research. Finally, some female scholars conducting fieldwork in their post-Soviet home country may face greater insecurities and risks than Western women due to the local gender dynamics (see Thibault's preface in this collection). In highlighting the discrepancies between local and nonlocal female researchers' position in the field, Wood, in her chapter, agrees with Central Asian female scholars (for example, Kudaibergenova 2019; Suyarkulova 2019), who questioned previous research that predominantly focused on the challenges faced by Western women conducting fieldwork in Central Asia. While nonlocal women who do not conform to local gender behavioral codes are only jokingly reprimanded for acting too much "like a man" or for being too "Western," as the accounts by white Western female scholars in this collection show, female researchers from the region are expected to conform to local gender roles.

In addition to physically impacting researchers' safety, fieldwork may also affect their mental health and well-being "regardless of where one's 'field' is" (Irgil et al. 2021, 15). As the personal story of Skriptaite demonstrates, her ability to conduct interviews from her apartment in the UK allowed research participants to enter and to observe her private life in ways that would not have been possible had they met offline. This meant that sometimes Skriptaite was unintentionally revealing more about herself to participants than she had intended. Not being able to control her image as a professional researcher in the field, left her in great emotional distress. In line with previous scholarship on fieldworkers' mental health, the personal accounts presented in this edited volume suggest a number of strategies to cope with emotional distress arising from fieldwork: taking breaks from the field, turning to informal support bubbles to "offload" or to debrief, counseling, exercise, and the use of a reflective diary (Mügge 2013; Loyle and Simoni 2017; Hummel and Kurd 2021; Irgil et al. 2021).

Peinhopf and Kamarauskaitė found that having breaks between research trips not only allowed them to reevaluate the collected data more critically and to review their own convictions about the field but also helped them reduce the negative effects of stress and anxiety related to their fieldwork experiences. It also helped them to maintain a healthy relationship with interlocutors and to stay focused while collecting and evaluating their data. In addition, taking breaks allowed Peinhopf to recuperate from her experience of betrayal. Yet, it also helped her to become less dependent on key informants. As she discusses in her chapter, it was through taking a step away from the field that she was able to understand that there will always be people who are willing to engage with her and that there was no need to keep in touch with specific participants with whom she no longer felt comfortable. Through commuting between Lithuania and the UK, Kamarauskaitė was likewise able to build an emotional, as well as physical, boundary between herself and her research participants. Moreover, since Kamarauskaitė was doing her field research "at home," she was able to recharge her batteries and discuss her anxieties with her family and close friends, without needing to leave the field. All this enabled her to be more reflective about her interactions and relationships with her interlocutors, with whom she shared an overlapping identity due to her sexual orientation. Drawing on their own fieldwork experiences, Peinhopf and Kamarauskaitė show us the value, in some instances, of leaving the field if a researcher is overwhelmed by homesickness and anxieties, as remaining can sometimes be counterproductive for the progression of the research project.

Navigating Ethical Dilemmas in the Field

The personal accounts of contributors to this volume outline some of the unique challenges that the post-Soviet region presents. When conducting research, scholars across a number of disciplines are expected to heed ethical guidelines to safeguard the privacy and well-being of themselves and their research participants. The guidelines are laid out by professional organizations such as the American Anthropological Association (American Anthropological Association 2012) and the American Political Science Association (APSA 2012). In line with previous scholarship on field research in the states of the former Soviet Union (De Soto and Dudwick 2000; Wall and Mollinga 2008; Roberts 2012; Gentile 2013; Lottholz 2018; Heathershaw and Mullojonov 2020), the authors in this collection argue that it is important to be mindful, to invest time in building rapport, and sometimes to radically change a research method in favor of locally acceptable modes of conduct to navigate some of those dilemmas in the field. The authors identify four broad dilemmas that early-career scholars and senior fieldworkers may encounter while doing research in the region: obtaining informed consent, ongoing consent, data storage/confidentiality, and role conflicts.

The first dilemma concerns the well-known issue of obtaining informed consent from respondents in the field. Previous research has suggested that in the more "closed" states of the former USSR, people are more reluctant to identify themselves during an interview, even with the assurance by the scholar that their statements will be treated anonymously (Rivera, Kozyreva and Sarovskii 2002; Allina-Pisano 2009; Roberts 2012; Heathershaw and Mullojonov 2020). Providing research participants with a formal letter of introduction outlining the aims of the research project and its intended outcome may thus not be sufficient to put people at ease and gain their trust. Navigating the dilemmas around obtaining respondents' written consent can be even more challenging for PhD field researchers, as Skriptaite's account demonstrates. While experts from Kazakhstan and Belarus generally agreed to be interviewed on Microsoft Teams and to provide written consent, Turkmen experts were more reluctant to participate in an online interview and to sign a consent form, out of the fear that their statements would be recorded by the Turkmen intelligence services. To accommodate her Turkmen respondents' safety concerns, backed by her University Research Ethics Committee, Skriptaite changed her original idea of obtaining respondents' written consent. Instead, she decided to record her participants' consent verbally on an audio tape if they declined to provide their written consent. In addition, Skriptaite constantly reassured her respondents that their identities remained confidential, would be presented only in aggregate or anonymous form, and would only be used in her PhD project. As she shows in her chapter, these measures helped Skriptaite to gain the trust of some of her interlocutors.

The second set of challenges outlined by the contributors to this collection concerns issues of "ongoing consent" (Wiles 2013, 74). Previous scholarship on field research suggests that informed consent requires an ongoing process of discussion, reflection, and negotiation of trust between the researcher and the research participant (Duncombe and Jessop 2012). Because scholars often build personal connections with their interlocuters when embedded in field sites for a long time, it can be tricky to maintain and navigate the boundaries between friendship and research participants' relationships, depending on the researchers' own position in the field (Tillmann-Healy 2003). Kamarauskaitė, Karas, and to a lesser extent Howlett note in their chapters that some relationships with their participants grew into friendships. To avoid any exploitation of trust or misuse of personal information, they constantly revaluated their role as researchers and friends during their fieldwork. Their personal stories accordingly highlight the importance of obtaining participants' "ongoing informed consent" before using certain information, which they might have shared with a scholar as friends and not as respondents.

Another ethical consideration highlighted in this book is researchers' obligations to properly handle and store the collected data, to conceal the identity of research participants after the interview. Post-Soviet regimes are known for using the Internet and new high-tech tools, such as artificial intelligence, CCTV cameras, "smart filtering," and hacking spyware to monitor their citizens' activities (Nazar 2018; Marat and Sutton 2021). Academics (Gentile 2013; Lottholz 2018) likewise can be the target of local intelligence services' wiretapping and hacking spyware and find their own and their respondents' privacy intruded upon. As malware such as Pegasus[5] and UK FinFisher has shown, data encryption does not prevent state agencies or other groups from accessing sensitive data. Accordingly, the transfer and storage of data collected in the field raises additional risks. Due to the risk of "dataveillance" (Lee 2019, 953), respondents might be more reluctant to participate in a virtual interview, as Skriptaite's story shows. Similarly, Turlubekova frequently reminded her research participants that she was collecting data for her study on drug trafficking in Kazakhstan. Knowing that she was required to report any relevant information on planned or committed criminal activities by her respondents to the local legal authorities if contacted, she paused the audio recording as soon as her interlocutors began sharing sensitive information that could have been used against them in court. Additionally, she included in her information leaflet a paragraph explicitly stating that data confidentiality would cease when requested to do so by the legal authorities in Kazakhstan. However, since most of the crime-relevant information provided by her interviewees was already public knowledge and had already been processed by the local criminal justice system, Turlubekova managed to navigate the ethical dilemmas around issues of data "confidentiality" in admirable fashion (Wiles 2013, 45). Moreover, because she was fully transparent about her role and obligations as a researcher, her participants seemingly felt more comfortable and willing to engage with her.

The fourth batch of challenges described by the contributions to this volume relates to a set of circumstances in which the authors' own value systems clashed with the views of their respondents (Kloos 1969). In such situations, previous literature suggests that scholars are often forced to hide their own feelings and opinions about a certain topic or issue out of the fear that speaking their mind may alienate their informants. These "role conflicts" (Kloos 1969, 509; Wiles 2013, 76) can cause considerable emotional stress for early-career field researchers, as the personal accounts presented here by Turlubekova, Howlett, and Karas suggest. Turning down dinner invitations as well as gifts offered by her participants, due to concerns about the associated expectations and her own personal safety, placed Howlett in a quandary. On the one hand, she had to follow her own safety and ethics protocol. On the other, Howlett was worried that her decision to reject Ukrainian hospitality could jeopardize her relationships with her interlocutors and harm her access to the field. In contrast to Howlett, Karas decided to follow local customs, keep her personal opinions to herself, and avoid asserting opposing opinions. Whilst back home in the UK with her friends, Karas would have felt compelled to counter the Sinophobia expressed by her participants, remaining silent as a participant observer and friend in the field helped her to maintain a positive relationship with her participants. Similarly to Karas and other Central Asian female scholars, Turlubekova soon noticed that the best method to empower her respondents was to take them "seriously, on their own terms" (Mamadshoeva 2019) and to listen carefully to understand how they make sense of the world of drug trafficking. In summary, the personal stories in this collection encourage other field researchers to critically reflect on how they interact with people holding different values, both in the field and at home.

Concluding Remarks

Given the rich detail and insightful reflections in this volume, it is my hope that this book will be a useful teaching tool in pre-fieldwork courses, prompting soon-to-be fieldworkers to think about the various challenges that they may (or may not) encounter before entering the field. Not every female researcher will experience the same ethical, personal, or methodological challenges as those outlined here. Still, the chapters offer insights into how those concerned adapted when facing adversity and unforeseen dilemmas. I hope this will aid other researchers in thinking about how they might also approach, cope with, and even avoid similar situations.

The personal stories presented in this collection also demonstrate that field research is an area not yet comprehensively addressed in conversations on diversity and equity in the broader academic profession, especially within the former Soviet space. It is the intersectionality of various categories of differences – age, education, sexuality, parental status, religious practices, class, and race – that not only shapes researchers' access to the field but also how they are received by their informants and the local society in the field. The reality beyond those categories of difference is that scholars' university affiliation, funding, and private financial means influence their access to the field and the progress of their research. While remote fieldwork has proven advantageous for some scholars amidst the COVID-19 pandemic – an aspect discussed by some of the contributors to this volume – there is a need to pay more attention to the growing financial and reputational inequality between researchers at different institutions, and the ways this affects their opportunities and experiences in the field. In light of the continuing uncertainties due to the COVID-19 pandemic and the growing political instability in the post-Soviet region (e.g. the war in Ukraine), I think these considerations are more important than ever. While I hope future volumes on field research practices will include an even greater variety of academic voices to highlight the difficulties faced by early-career scholars of various social and institutional backgrounds, this volume is just the beginning of a much-needed conversation.

Notes

1 To account for the various forms of field research, including remote methods, the collection defines fieldwork as conceptualized by Irgil et al. (2021, 6). He stated that fieldwork involves acquiring information, using appropriate data collection methods, for qualitative, quantitative, or experimental analysis through embedded research whose location and duration are dependent on the research project.

2 As lead editor of this volume, I agree with England (1994, 84) that field research is a by-product of the positionality of both participant and researcher. Being reflexive therefore allows scholars to identify how their own position influences fieldwork progress and knowledge production. The term "intersectionality" was coined by Crenshaw (1991, 1244) to describe the different "ways in which race and gender interact to shape the multiple dimensions of black women's employment experiences." Since then, the concept's use has extended to other marginalized groups, which is why "intersectionality" in this book refers to the multiple positions and identities (e.g., class, sexuality, race, and nationality) that researchers hold and that inform their inclusion, exclusion, and access to the field (Gill 2007).

3 I define remote fieldwork as "the collection of data over the Internet or over the phone" (Irgil et al. 2021, 19).

4 Following the collapse of the Soviet Union, homosexuality was legalized in almost all post-Soviet states, apart from Uzbekistan and Turkmenistan, where same-sex relationships remain illegal to this day (Mole 2019). The collapse of the USSR in 1991 gave rise to hopes of greater tolerance toward non-heterosexuals. Yet, post-Soviet society, including state officials and religious authorities, continues to be homophobic and transphobic. At the state level, acceptance of homosexuality and gender variance has been framed as being in direct contradiction to the physical and moral well-being of the nation, with governments arguing for the necessity of protecting the public, and especially young people, from morally harmful information that may lead them astray and prevent them from fulfilling their heterosexual duty to the nation. Such concerns have led authorities in several countries to limit minors' access to information about sexual and gender diversity via proposals for so-called anti-gay propaganda bills aimed at protecting "traditional" – e.g., heteropatriarchal and reproductive – relationship norms (Wilkinson 2020).

5 Pegasus is a cutting-edge spyware product made by the Israeli company NSO Group. It infects iPhones and Android devices to enable operators of the tool to extract messages, photos, and emails, record calls and secretly activate microphones. So far, there is a documentary evidence that at least ten regimes use Pegasus: Azerbaijan, Bahrain, Kazakhstan, Mexico, Morocco, Rwanda, Saudi Arabia, Hungary, India, and the United Arab Emirates. In addition to business executives, religious figures, NGO employees, union officials, government officials, including presidents, and prime ministers, academics have also been targeted by the malware (Kirchgaessner et al. 2021).

References

Adams, L. 1999. "The Mascot Researcher. Identity, Power, and Knowledge in Fieldwork." *Journal of Contemporary Ethnography* 28 (4): 331–363. DOI: 10.1177/089124199129023479.

Allina-Pisano, J. 2009. "How to Tell an Axe Murderer: An Essay on Ethnography, Truth, and Lies." In *Political Ethnography: What Immersion Contributes to the Study of Power*, edited by E. Schatz, 53–73. Chicago: University of Chicago Press.

American Anthropological Association. 2012. *Principles of Professional Responsibility*. Available at: https://www.americananthro.org/LearnAndTeach/Content.aspx-?ItemNumber=22869 (accessed on March 28, 2022).

American Political Science Association. 2012. *A Guide to Professional in Political Science*. 2nd edition. Washington: American Political Science Association. Available at: https://www.apsanet.org/portals/54/Files/Publications/APSAEthicsGuide2012.pdf (accessed on March 28, 2022).

Crenshaw, K. 1991. "Mapping the Margins: Intersectionality, Identity Politics, and Violence Against Women of Color." *Stanford Law Review* 43 (6): 1241–1299. DOI: 10.2307/1229039.

Cromwell, A. and M. Tadevosyan. 2021. "Deconstructing Positionality in Conflict Resolution: Reflections from First-Person Action Research in Pakistan and the South Caucasus." *Action Research* 19 (1): 37–55. DOI: 10.1177/1476750320960803.

Dall'Agnola, J. 2020. "Queer Culture and Tolerance in Kazakhstan." In *PC on Earth. The Beginnings of the Totalitarian Mindset*, edited by J. Dall'Agnola and J. Moradi, 99–116. Stuttgart, New York: ibidem, Columbia University Press.

Dall'Agnola, J. 2021. *The Impact of Globalization on National Identities in Post-Soviet Societies*. PhD dissertation, Oxford Brookes University.

Dall'Agnola, J. 2022. "'Tell Me Sister' - Social Media, a Tool for Women Activists in Tajikistan." *Central Asian Affairs* 9 (1): 119–147. DOI: 10.30965/22142290-12340018.

Dekeyser, T. and B. Garret. 2021. "Illegal Ethnographies. Research Ethics beyond the Law." In *Research Ethics in Human Geography*, edited by S. Henn, J. Miggelbrink and K. Hörschelmann, 153–167. London: Routledge. DOI: 10.4324/9780429507366-9.

De Soto, H. and N. Dudwick. 2000. *Fieldwork Dilemmas. Anthropologists in Postsocialist States*. Wisconsin: The University of Wisconsin Press.

Duncombe, J. and J. Jessop. 2012. "'Doing Rapport' and the Ethics of 'Faking Friendship'." In *Ethics in Qualitative Research*, edited by T. Miller, M. Birch, M. Mauthner and J. Jessop, 108–121. London: SAGE Publications Ltd. DOI: 10.4135/9781473913912.

England, K. 1994. "Getting Personal: Reflexivity, Positionality, and Feminist Research." *Professional Geographer* 46 (1): 80–89. DOI: 10.1111/j.0033-0124.1994.00080.x.

Gentile, M. 2013. "Meeting the 'Organs': The Tacit Dilemma of Field Research in Authoritarian States." *Area* 45 (4): 426–432. DOI: 10.1111/area.12030.

Gill, V. 2007. "Theorizing and Researching Intersectionality: A Challenge for Feminist Geography." *The Professional Geographer* 59 (1): 10–21. DOI: 10.1111/j.1467-9272.2007.00587.x.

Heathershaw, J. and P. Mullojonov. 2020. "The Politics and Ethics of Fieldwork in Post-Conflict Environments: The Dilemmas of a Vocational Approach." In *Doing Fieldwork in Areas of International Intervention: A Guide to Research in Violent and Closed Contexts*, edited by B. Bliesemann de Guevara and M. Bøås, 93–112. Bristol: Bristol University Press. DOI: 10.46692/9781529206913.007.

Howlett, M. 2021. "Looking at the 'Field' through a Zoom Lens: Methodological Reflections on Conducting Online Research during a Global Pandemic." *Qualitative Research* (January): 1–16. DOI: 10.1177/1468794120985691.

Hummel, C. and D. Kurd. 2021. "Mental Health and Fieldwork." *P.S.: Political Science & Politics* 54 (1): 121–125. DOI: 10.1017/S1049096520001055.

ILGA-Europe. 2022. "Annual Review of the Human Rights Situation of Lesbian, Gay, Bisexual, Trans and Intersex People in Europe and Central Asia." Available at: https://rainbow-europe.org/annual-review (accessed on March 28, 2022).

Irgil, E., A. Kreft, M. Lee, C. Willis and K. Zvobgo. 2021. "Field Research: A Graduate Student's Guide." *International Studies Review* (June 2021): 1–23. DOI: 10.1093/isr/viab023.

Johansson, L. 2015. "Dangerous Liaisons: Risk, Positionality and Power in Women's Anthropological Fieldwork." *Journal of the Anthropological Society of Oxford* 7 (1): 55–63.

Johnson, J. 2009. "Unwilling Participant Observation among Russian *Siloviki* and the Good-Enough Field Researcher." *PS: Political Science & Politics* 42 (2): 321–324. DOI: 10.1017/S1049096509090647.

Kirchgaessner, S., P. Lewis, D. Pegg, S. Cutler, N. Lakhani and M. Safi. 2021. "Revealed: Leak Uncovers Global Abuse of Cyber-Surveillance Weapon." *The Guardian*, July 18. https://www.theguardian.com/world/2021/jul/18/revealed-leak-uncovers-global-abuse-of-cyber-surveillance-weapon-nso-group-pegasus?CMP=Share_iOSApp_Other.

Kloos, P. 1969. "Role Conflicts in Social Fieldwork." *Current Anthropology* 10 (5): 509–512. DOI: 10.1086/201052.

Kudaibergenova, D. 2019. "When Your Field Is Also Your Home: Introducing Feminist Subjectivities in Central Asia." *openDemocracy*, October 7. https://www.opendemocracy.net/en/odr/when-your-field-also-your-home-introducing-feminist-subjectivities-central-asia/.

Lee, C. 2019. "Datafication, Dataveillance, and the Social Credit System as China's New Normal." *Online Information Review* 43 (6): 952–970. DOI: 10.1108/OIR-08-2018-0231.

Lottholz, P. 2018. "Researcher Safety in Peace, Conflict and Security Studies in Central Asia and Beyond: Making Sense and Finding New Ways Forward." *Security Praxis*: 1–7.

Loyle, C. and A. Simoni. 2017. "Researching under Fire: Political Science and Researcher Trauma." *PS: Political Science & Politics* 50 (1): 141–145. DOI: 10.1017/S1049096516002328.

Mamadshoeva, D. 2019. "Listening to Women's Stories: The Ambivalent Role of Feminist Research in Central Asia." *openDemocracy*, October 9. https://www.opendemocracy.net/en/odr/listening-to-womens-stories-the-ambivalent-role-of-feminist-research-in-central-asia/.

Marat, E. and D. Sutton. 2021. "Technological Solutions for Complex Problems: Emerging Electronic Surveillance Regimes in Eurasian Cities." *Europe-Asia Studies* 73 (1): 243–267. DOI: 10.1080/09668136.2020.1832965.

Milligan, L. 2016. "Insider-Outsider-Inbetweener? Researcher Positioning, Participative Methods and Cross-Cultural Educational Research." *Compare: A Journal of Comparative and International Education* 46 (2): 235–250. DOI: 10.1080/03057925.2014.928510.

Mole, R. 2019. "Constructing Soviet and Post-Soviet Sexualities." In *Soviet and Post-Soviet Sexualities*, edited by R. Mole, 1–15. London: Routledge.

Mügge, L. 2013. "Sexually Harassed by Gatekeepers: Reflections on Fieldwork in Surinam and Turkey." *International Journal of Social Research Methodology* 16 (6): 541–546. DOI: 10.1080/13645579.2013.823279.

Nazar, N. 2018. "How Turkmenistan Spies on Its Citizens at Home and Abroad." *openDemocracy*, August 16. https://www.opendemocracy.net/en/odr/how-turkmenistan-spies-on-its-citizens/.

Rivera, S., P. Kozyreva and E. Sarovskii. 2002. "Interviewing Political Elites: Lessons from Russia." *PS: Political Science & Politics* 35 (4): 683–688. DOI: 10.1017/S1049096502001178.

Roberts, S. 2012. "Research in Challenging Environments: The Case of Russia's 'Managed Democracy'." *Qualitative Research* 13 (3): 337–351. DOI: 10.1177/1468794112451039.

St. Julian-Varnon, K. 2020. "A Voice from the Slavic Studies Edge: On Being a Black Woman in the Field." *NewsNet ASEEES* 60 (4): 1–4. https://www.aseees.org/news-events/aseees-blog-feed/voice-slavic-studies-edge-being-black-woman-field.

Suyarkulova, M. 2019. "A View from the Margins: Alienation and Accountability in Central Asian Studies." *Open Democracy*, October 10. https://www.opendemocracy.net/en/odr/view-margins-alienation-and-accountability-central-asian-studies/.

Tadevosyan, M. 2019. *Multidimensional Roles of Local Non-Governmental Organizations in Creating Reconciliation Spaces in the South Caucasus*. PhD dissertation, George Mason University.

Thibault, H. 2021. "'Are You Married?': Gender and Faith in Political Ethnographic Research." *Journal of Contemporary Ethnography* 50 (3): 395–416. DOI: 10.1177/0891241620986852.

Tillmann-Healy, L. 2003. "Friendship as Method." *Qualitative Inquiry* 9 (5): 729–749. DOI: 10.1177/1077800403254894.

Wall, C. and P. Mollinga. 2008. *Fieldwork in Difficult Environments. Methodology as Boundary Work in Development Research*. Berlin: LIT Verlag.

Wiles, R. 2013. *What Are Qualitative Research Ethics?* London: Bloomsbury Collections.

Wilkinson, C. 2020. "LGBT Rights in the Former Soviet Union: The Evolution of Hypervisibility." In *The Oxford Handbook of Global LGBT and Sexual Diversity Politics*, edited by M. Bosia, S. McEvoy, and M. Rahman. Oxford: Oxford University Press. DOI: 10.1093/oxfordhb/9780190673741.013.12.

Part I

Stories from the Post-Soviet Field

1 Understanding and Managing One's Own Mistrust

The Value of Embodied Ethnography during Fieldwork in a Contested Postwar Polity[1]

Andrea Peinhopf
University of York

Introduction

Following the dissolution of the Soviet Union in 1991, violent ethnic conflict broke out in several post-Soviet states, including Moldova, Georgia, and Azerbaijan. The wars over Transnistria, Abkhazia, South Ossetia, and Nagorno-Karabakh led to the long-term displacement of hundreds of thousands of civilians and the establishment of so-called *de facto* states. While long dismissed as "black holes" (e.g., Lynch 2007, 486), or Russian puppet states with little agency worth exploring on its own, the post-Soviet unrecognized states have become increasingly popular fieldwork destinations for Western researchers over the last decade. Political and other social scientists, including anthropologists and sociologists, have provided new insights into the internal dynamics of these often-shunned entities (e.g., Chamberlain-Creanga 2008; Clogg 2008; Blakkisrud and Kolstø 2012; Kabachnik 2012; Shahnazarian and Ziemer 2014; Peinhopf 2021). However, while there is now a growing body of literature dealing with the various facets of *de facto* statehood and unresolved conflict, little is said about the process of data collection itself and the particular challenges that researchers can encounter in places marked by unresolved war and international isolation.[2]

In this chapter, I take the opportunity to reflect on my personal experience conducting long-term ethnographic fieldwork in Abkhazia between 2016 and 2018 on the topic of postwar identification. More specifically, the chapter focuses on my positionality as a Western researcher and young woman and the feelings of mistrust that these positions could trigger vis-à-vis certain contacts in the field. In recent years, more attention has been paid to how fieldwork, and hence the data one can collect, is shaped by "who the researcher is, and is allowed to be" (Krause 2021, 330). With this reflexive turn, it is no longer assumed that researchers are detached, neutral observers and that everyone can access the same kind of data regardless of their age, gender, race, and sexuality, among other factors. Nevertheless, these ideas are far from mainstream, especially in traditionally positivistic

DOI: 10.4324/9781003144168-3

disciplines such as political science, where the notion of the neutral observer continues to prevail. Focusing on the issue of mistrust *from the perspective of the researcher*, the chapter therefore moves beyond the predominant emphasis on how to gain local peoples' trust and build rapport. Instead of simply investigating mistrust vis-à-vis myself as a Western researcher and outsider, the chapter explores my own suspicions in relation to others as a by-product of being drawn into local power dynamics. It thus foregrounds the power that research *participants* can hold over researchers, as opposed to a more common focus on the position of power of the researcher and the duty to protect research participants from potential harm (e.g., Wood 2006; Gallaher 2009).

The chapter accordingly highlights some of the challenges associated with conducting immersive research in a conflict-ridden place like Abkhazia; however, it does so not with the intent of discouraging anyone (and especially young women) from conducting fieldwork. Instead, it outlines a way in which difficult experiences can enrich the research process and even be written into the text. Drawing on the concept of embodied ethnography (Hanson and Richards 2019), it suggests that certain challenges in the process of data collection are best understood not as *limits to* data collection but as *part of* the data. Such an approach can be particularly useful for women working in difficult settings, but it is also beneficial for male researchers; as Rebecca Hanson and Patricia Richards have noted, "men must also be able to speak openly about threats and risks they face in the field [...] without being criticized for not meeting the hegemonic standards of masculinity upheld by ethnography" (2019, 177). Moreover, while the main focus of the chapter is on ethnographic research in the Abkhazian context, it hopes to provide insights for those conducting short-term, interview-based fieldwork as well as for those working in or on other unrecognized states and conflict-affected regions, both within and beyond the former Soviet Union.

The remainder of this chapter is organized as follows. In the first section, I provide a brief overview of the aims and practicalities of my fieldwork and reflect on my positionality as a Western researcher and young woman. This is followed by a longer section detailing how this positionality played out in the process of field research. It centers on interactions with one of my key interlocutors, who staged his own disappearance, and the pervasive mistrust and paranoia that this caused. I then demonstrate how it was precisely the experience of profound mistrust that deepened my understanding of Abkhazian postwar society and, paradoxically, brought me closer to an insider position. This particular incident thus highlights the value of an embodied approach, as it allowed me to reconceptualize my experience not as a failure to conduct "proper" fieldwork but as an essential part of the fieldwork. This not only required a more reflexive stance but also required for me to rethink my experience in the context of local power dynamics. In the conclusion, I reflect on what I learned from this experience and what I wish I had done differently. In doing so, I underline the significance of an "embodied reflexivity" (Hanson and Richards 2019, 3) and a more flexible approach to ethnography that encourages "limited" or "uneven immersion" (Krause 2021, 330–335) where necessary.

Conducting Difficult Fieldwork as a Western Researcher and Young Woman

The Georgian-Abkhaz war was one of the bloodiest in the former Soviet Union. It began on August 14, 1992, when (para)military groups sent by the central government in Tbilisi entered Abkhazia, and ended 13 months later with the mass displacement of over 200,000 Georgians, most of whom have not been able to return (Human Rights Watch 1995). What were the causes and consequences of this intercommunal violence? Although there is considerable interest in the post-Soviet "frozen conflicts," few studies have taken a close look at the violence out of which most of them emerged; there seems to be an implicit assumption that warfare was essentially stirred and executed not by the actors on the ground but by an outside power (Russia). Adopting a people-centered approach, my PhD thesis, in contrast, sought to uncover the intimate stories behind the violence – why it broke out, what it did to people, and how it continues to shape interpersonal relationships.

To explore the causes and consequences of intercommunal violence, I decided to pursue long-term ethnographic research, combining life-history interviews and participant observation. In order to do so, I first needed to apply for ethics approval at my university and complete a thorough risk assessment. Contrary to popular belief, the Georgian government as well as the *de facto* Abkhazian authorities allow foreigners to cross the *de facto* Georgian-Abkhazian border, officially called the Administrative Boundary Line (ABL) in Georgia proper. As Georgia continues to see Abkhazia as part of its territory, there is no official "border" checkpoint. Instead, one is required to show one's passport at a small police station, where officials take a copy. To enter the Abkhazian-controlled territory, one needs to apply for an entry permit in advance (which can be done online), followed by a visa application upon entry (Ministry of Foreign Affairs of the Republic of Abkhazia 2022).

While traveling to Abkhazia via Georgia is not illegal, it does go against the official advice of the UK's Foreign, Commonwealth and Development Office (FCDO).[3] The main reasons are security concerns, most importantly, that there is no consular service available (GOV.UK 2022). However, when I started my PhD, there had been no significant fighting for more than a decade and the political situation within Abkhazia was stable.[4] It was a time when a growing number of Western scholars had conducted field-work in Abkhazia, in particular on topics around *de facto* state-building and democratization (e.g., Ó Beacháin 2012; Kolstø and Blakkisrud 2013). While most of these studies relied on short-term fieldwork, there were also a number of publications based on long-term, ethnographic field research (e.g., Shesterinina 2014; Costello 2015). These sources in combination with a previous visit as a tourist helped me to gain a more nuanced understanding of the security situation on the ground. In addition, I also consulted with international organizations and their local branches. Based on this information, I identified the locations that could be considered safe for research. Following FCDO advice, I stayed away from the areas immediately surrounding the ABL. On the condition that potential risks would be assessed by international organizations on an ongoing basis, my fieldwork was authorized by my university.

Starting in 2016, I conducted a total of eight months of fieldwork over a period of two years. Entering Abkhazia through Georgia meant that I was only granted shorter, single-entry visas from the *de facto* authorities, which forced me to plan a number of shorter visits (usually two months at a time) instead of one long stay.[5] Inside Abkhazia, I spent most of my time in and around the capital of Sukhum as well as in a village in the eastern Ochamchira district, which had a mixed Georgian-Abkhaz population before the war and is now predominantly Abkhaz.[6] In addition, I traveled around Abkhazia as much as I could, visiting people I met in Sukhum who then introduced me to their extended families in the village. Overall, 45 people participated in my research, out of which 10 were so-called "key informants," with whom I was in regular contact. Among them were farmers, teachers, petty traders, racketeers, taxi drivers, and nurses (i.e., they were non-elite actors). Although I was expected to bond with local women of a similar age, my encounters were mixed in terms of gender. Regarding ethnicity and age, my interlocutors were predominantly – but not exclusively – ethnic Abkhazians who still remember peaceful prewar cohabitation, while also having had firsthand experience of the violence during the war. My aim was to immerse myself in their everyday lives in order to understand how they make sense of the war and postwar period, including the mass displacement of the local Georgians as well as Abkhazia's international isolation as a punitive response to it.

There were many factors that shaped my access and interactions with people in the field. First of all, as a researcher from the West, I was exposed to certain levels of resentment and mistrust. This had to do with the issue of international nonrecognition and the heightened sense of threat and "siege mentality" that it had fostered (Caspersen 2012, 34; see also Bryant 2014, 128). As someone who was born and raised in Austria and living in the UK, I was not in principle seen as involved in the conflict. I was nevertheless from an area of the world that had not only failed to recognize Abkhazia's independence but also ignored the grievances of its people more generally. As a consequence, I was viewed as somewhat complicit in the international isolation that Abkhazia has faced. My decision to enter Abkhazia through Georgia exacerbated this perceived complicity. While I learned to appreciate my journeys from "Georgia proper" for the insights they provided into the complexities of the conflict, it was precisely this exposure to multiple perspectives that was seen as problematic in Abkhazia.

In fact, very early in my fieldwork, one of my gatekeepers reprimanded me for supposedly not being aware enough of the negative consequences that my presence had for my contacts' reputation if I decided to continue entering through Georgia. While this was certainly well-meant advice, it also had a threatening undertone, and, as a consequence, strengthened my decision to work (and live) as independently as possible. That I had neither an official affiliation nor a research assistant gave me some independence and peace of mind, but it also made me reliant on people's goodwill and hospitality for accessing certain locations. This would not have been a big problem had I not been a young, unmarried woman. Working independently meant that I was often on my own. While this was a normal – and even expected – behavior for a researcher (especially an ethnographer), it was not in line with the local norms for women's behavior. Although alternative standards applied to foreign women, they were a double-edged sword; despite giving me some freedom in terms of movement and interaction with a wide range of people, they also carried certain risks, as they were largely based on the stereotype of the Russian female tourist looking for a holiday romance. To some extent, these alternative standards therefore came with the expectation that foreign women were more open and available sexually. In addition, Western women were also associated with wealth and material privilege, which could make them even more attractive. The position of a young Western woman therefore exposed me to different challenges than that of the Western researcher. While the latter made me susceptible to hostility and suspicion, the former could elicit sexualized interactions ranging from aggressive flirting to sexual harassment.[7]

The combination of these two positionalities hence created a particular sense of vulnerability that became more and more pronounced in the course of my field research. While it manifested itself on numerous occasions, the next section is focused on a particular experience with one male interlocutor that both caused much frustration and proved to be particularly insightful for my research. In doing so, I highlight the so-called "awkward surplus" (Fujimura 2006, 51) or data that do not fit into preestablished categories of what counts as data and are therefore often considered residual or even ignored altogether. In the context of ethnographic research, this can include certain challenges and dilemmas, in particular the more uncomfortable and frustrating experiences that are usually seen as more appropriate for so-called "venting journals" than serious academic writing, often out of concerns about one's professional reputation and career, which can be even greater among early-career scholars (see Hanson and Richards 2019, 154–164).[8]

The "Awkward Surplus:" Betrayal and Mistrust in the Field

When I was traveling from a small urban settlement to Sukhum very early in my fieldwork, a young Abkhaz man (Rustam) sat down next to me with a bag of tangerines.[9] After only a few minutes, he smiled at me and said:

"I can see you are not from here. Have you tried our famous tangerines before?" I accepted the offer, and he started telling me about the places we were passing as we drove along the main road toward the capital. Before I got out on the outskirts of Sukhum, he gave me his phone number and invited me to visit him at his workplace at a farm in one of the villages. Having told him about my research, he was also eager to put me in touch with a local journalist, who, he claimed, would be happy to share his personal knowledge about the conflict.

Back in Sukhum, I was uncertain whether to follow up on the offer. As someone who had grown up in postwar Abkhazia and lived in a village, Rustam was an interesting and useful contact. Moreover, he appeared to be an open and generous person. I was nevertheless worried about the potential romantic or sexualized nature of the invitation, asking myself whether I had made it clear enough that I was in Abkhazia as a researcher. However, after a few days, I decided to call him. Following the advice from my fieldwork training and the broader fieldwork literature, I explained that I wanted to meet on the condition that he would tell me his full name and address, as well as that of the journalist, so that I could pass the information to the Abkhaz friend I was staying with as well as to contacts in the UK. He was amused by my precautions but gave me all the information I asked for. I also reiterated my research interests to underline my professional identity. The next morning, wearing my most functional clothes, I took an early *marshrutka* (minibus) to the village where he worked. After picking me up in his car, he first showed me around the farm and then drove to the journalist's house, where we were offered homemade food and wine. The atmosphere was friendly, and the elderly journalist was keen to share his knowledge.

In the late afternoon, Rustam drove me back to the bus stop. Before I got out of the car, he said that he wanted to ask me something. After hesitating for a moment, he told me how much he had enjoyed our time together, told me that he had never met a woman like me before, and, out of the blue, asked if I could imagine marrying him. Completely taken by surprise, I said that I also enjoyed his company but that I was not able to marry anyone involved in my research for ethical and professional reasons. Once on the bus, I felt deeply frustrated. Was it not possible to contact a man without receiving a marriage proposal? Or was this even my own fault because I decided not to wear a fake wedding ring, as it is often recommended?[10] By declining his offer, had I offended his male ego and lost him as a contact?[11] To my surprise, Rustam called only a few days later and explained that he, of course, understood and respected my position. After that, we stayed in touch and regularly chatted over the phone. As he seemed genuinely interested in my work, I enjoyed our conversations and benefitted from his advice and personal insights, which helped me to better understand the social reality of a young man growing up in an unrecognized, postwar polity. Given how open he was, I felt that I could trust him and even share sensitive information about myself, including my journeys through Georgia proper.

Then, one day, he called me in great distress, explaining that his mother was very ill and needed surgery. Given his precarious economic situation and the urgency of the matter, he said that he had no other choice than to ask me for money. I was surprised and told him that my university did not allow me to lend money to contacts in the field. However, after hanging up, I felt conflicted, first, because of all the insights he had shared with me for my research and, second, because of a heightened sense of my own privilege – as what he had asked for was not a significant sum for me. At the same time, though, the prospect of lending him money made me deeply uncomfortable. I was not only concerned about transgressing professional boundaries and switching from the role of a researcher to that of a friend but was also equally worried about the impact that money could have on our personal relationship.[12] Eventually, after discussing the situation with a colleague with experience in the region, I came up with a compromise. I called him back to explain that while I was not allowed to lend money, I had a certain budget for a driver if he was willing to help me arrange some research trips to more remote places. He agreed, and a few days later, I gave him the money as an advance payment. This was the last time I would see or hear from him in a long time.

At first, I did not think much of it. But when he did not answer my calls, I began to wonder. One day, a woman picked up my call and told me that he had left his phone and simply vanished. She later sent me several text messages, explaining that the family was getting increasingly worried and looking for him everywhere. They were afraid something had happened to him. As the news was distressing, I did not know what to do or think. Was it really possible that he had simply disappeared? And who should I talk to? I decided to reach out to an NGO employee near the region where he was allegedly last seen. Reluctantly, I sent my contact a text message, asking whether he had heard of anyone who had disappeared in his area. He quickly replied saying that he was not aware of any such case and that the story sounded suspicious. When I explained the situation in greater detail, he had no doubt that Rustam's disappearance was fabricated: he got the money and that was all he needed.

According to the NGO employee, who was ethnically Georgian, "that's what they do – they screw you over." Given his prejudice vis-à-vis ethnic Abkhazians, I was hesitant to take his words at face value. Knowing that I should speak with an Abkhaz person, I decided to confide in my neighbor, Zhanna, a very caring middle-aged woman who often invited me for coffee. When I visited her a few days later, I carefully outlined the situation and asked about the likelihood of the scenario and whether she had recently heard of anyone who had gone missing. First, she was hesitant, saying the story was theoretically possible, but given Abkhazia's small size, she would have certainly heard about it. She then picked up her phone and called a friend living in the same village as Rustam's family. Like herself, this person did not know of any case of a missing person. As Zhanna was becoming increasingly suspicious, she asked me for Rustam's phone number. After ringing a few times, somebody answered and Zhanna quickly switched from Russian to Abkhaz.

To my amazement, it took her only a few minutes to get hold of the allegedly missing Rustam. While I could not completely follow the conversation, I certainly understood its tone as Zhanna became louder and louder.[13] When she hung up, she exasperatedly exclaimed "*Svoloch*!" (Russian for "bastard"). She explained that Rustam had not gone missing, nor was his mother in need of surgery. However, he was not planning to return the money because he maintained it had been a gift; he claimed we had met online and that I simply offered to send him money when he shared his problems with me. Also, "we were not married," so I could not demand anything from him. When Zhanna threatened that I would go to the police, he told her that I was free to do so, but he would have to tell the police "what I was really doing here," suggesting I was a spy working for the Georgian government. Finally, after insisting he was not going to argue with a woman, he simply hung up. Zhanna was frantic but also aware there was little to be done. She concluded: "Rustam knows that you are a foreigner and is taking advantage of that."

Shifting Positionalities: Mistrust as a Marker of Local Belonging

Although it was a relief to know the truth about Rustam's alleged disappearance, the incident took a toll on me. First and foremost, it raised certain questions: had he targeted me from the beginning, or did certain circumstances arise that, in his view, justified this kind of betrayal? I realized that as a young Western woman, I was a popular target for so-called *aferisty* (Russian for "swindlers"), both because of my lack of in-depth cultural intimacy (especially at the early stages of my fieldwork) and the material privilege associated with people from Western Europe. While men would not necessarily be excluded from such betrayal, Rustam had taken advantage of the constraints and risks that I faced as a young female researcher working alone in the field, drawing on the gendered role of the "male protector" who would ensure my safety and provide me with access to places that I might not otherwise be able to visit.

But, as mentioned earlier, my identity as a Western researcher, particularly my association with Georgia and the West, also proved problematic in this situation as it meant that my credibility could be easily challenged. That I had no local affiliation or formal research assistance made it even more difficult for me to defend myself. In addition, the fact that neither my country of residence (UK) nor my country of origin (Austria) recognized Abkhazia as an independent state meant that I had little political clout as a foreign citizen. Rustam hence benefitted from the often exaggerated, yet partly accurate, notion of *de facto* states as lawless "black holes," which was particularly evident in his reaction to Zhanna's threat to go to the police. While his indifference was partly grounded on the weakness of Abkhazia's law enforcement (which can also be the case in a recognized state), he knew that the likelihood that I would receive assistance was low due to my status as a foreigner coming from a country with no diplomatic connections to Abkhazia.

Overall, the incident profoundly changed the way I looked at the people around me, turning the remainder of my fieldwork into an ongoing exercise in locating trust. In addition, I was overcome with self-doubt and routinely asked myself how this could have happened. Should I have known not to trust this person? Was I not suited to work as an ethnographer? In addition to my worry about "letting" this happen in the first place, and how it reflected my suitability as an ethnographer, I was also concerned about the consequences for my ability to continue my ethnographic investigation. At least in the immediate aftermath of the incident with Rustam, I was sometimes overcome with a sense of mistrust and paranoia, making it difficult to summon up the openness and curiosity that ethnographic research requires.

Yet, one afternoon when I was visiting a Russian couple that I often spoke with, my outlook changed profoundly. As we were enjoying coffee, I decided to share my misery with Yura, who always took great interest in my work. At first, he seemed very annoyed, asserting, "I told you not to trust him!" But his voice quickly softened as he revealed that they also had to learn all this the hard way: "Because there were no Abkhazians in our town before the war, or at least very few, we had no idea what their mentality was like. This is something we had to get used to ourselves after the war." He explained that there was much they had learned by making mistakes such as trusting the wrong people or expressing their opinion too openly; these negative experiences left them with intimate knowledge that became the basis for a sense of mistrust vis-à-vis ethnic Abkhazians. According to Yura, for an Abkhaz person, the day was spent in vain if (s)he did not betray or steal from anyone, as this is "something they take pride in. And there is no limit. They even betray each other. It is only within the family that they stick together." With this, he added that it is "best to keep your distance, smile, say 'thank you' and that everything is great. But do not let them take advantage of you."

Without taking his prejudice at face value, Yura's reaction helped me realize the extent to which mistrust is embedded in the social structures in Abkhazia, and how, as a consequence, my experience of betrayal simultaneously marked me as an outsider and an insider, thus blurring the boundaries between the two.[14] On the one hand, it highlighted a certain degree of naivety that was typical for an "outsider," and which made me vulnerable to being taken advantage of. As mentioned earlier, this vulnerability was not necessarily limited to foreign women, although they are certainly popular targets, especially during the summer months. In fact, there is a stark distinction between female visitors – mostly from Russia – and "our girls," referring to ethnic Abkhazians or, sometimes, women raised in Abkhazia more generally. Whereas "our girls" were for serious courtship and marriage, Russian women were there for "fun," making foreign women popular targets for so-called *aferisty*. On the other hand, by virtue of becoming a victim and "learning my lessons," I also came closer to being an insider. Unlike a "true outsider," I did not simply leave but had to stay and learn how to navigate society in the light of my experience. In this sense, my position resembled that of local Russians like Yura and his family, who had stayed after the war and had to adapt to a new postwar reality.[15]

Still, mistrust could not simply be reduced to an interethnic phenomenon. Only a few weeks after the incident, my Abkhaz friend, Khibla, invited me to come to the village with her for the weekend, explaining that she was tired of the people around her and needed a break. According to Khibla, there are so many people in Abkhazia who want to be friends with you or call themselves your "friend" just to take advantage of you, which is why she stressed that one "has to be very careful" and keep people at a distance. As my fieldwork unfolded, I increasingly realized that since the war had ended, it had become difficult for people in Abkhazia to trust *anyone*. Relations among people – even if they were of the same ethnicity – seemed constantly fraught with conflict. As one of my long-term contacts, an Abkhaz man in his late 30s, told me toward the end of my fieldwork, "you as a foreigner is one thing, but even we, who are all connected with each other through shared acquaintances or friends, live in constant fear of betrayal. *People have changed significantly*." Focusing solely on intergroup relations, prior studies (e.g., Clogg 2008; Trier, Lohm and Szakonyi 2010; Mühlfried 2019) have therefore failed to capture the extent to which Abkhazia as a whole can be seen as a "community of mistrust," as this wariness seemed a feature of almost *any* social interaction, whether between members of an ethnic group or across them. Although cultural intimacy without doubt existed among co-ethnics, being among "one's own" was not necessarily the safe space one may expect.

Rethinking What Counts as Data: The Value of Embodied Ethnography

Reflecting on the role of mistrust in Abkhazian society provided a fruitful angle to reevaluate my own fieldwork experiences. In particular, it raised questions around what counts as success (or failure) when conducting an ethnography. If betrayal (or the fear thereof) is a pervasive phenomenon, can my experience not be seen as part of the immersion and intimacy that is at the heart of ethnography? Some people suggested that I had simply met the "wrong" people, but, if betrayal is socially endemic, would I not missed out on a fundamental aspect of Abkhazian society had I socialized with the "right" ones? One could argue that by going throug experience with Rustam, I moved away from participant observation became an *observant participant* instead.[16] Ideally, I would have wanted to avoid experiencing betrayal myself and instead observe it happening around me. This, however, presupposes a degree of detachment that is not only unrealistic but also inherently problematic.

As Rebecca Hanson and Patricia Richards have suggested, there tends to be an expectation among researchers when entering the field that they would be first and foremost treated as professionals standing above local power relations and gender norms, an expectation that "reflects the colonialist legacy of ethnography and the assumption that researchers can somehow stand above and beyond the community they study" (2019, 21). This assumption is fostered by an institutional emphasis in fieldwork training and ethics committees on the power that researchers hold over their participants and the need to ensure the participants' safety and well-being. However, as Carolyn Gallaher has pointed out, "vulnerability is not exclusive to research subjects" (2009, 135). While undoubtedly crucial, focusing solely on the rights of those being researched overlooks the power that participants can also hold over researchers, especially when they are at the early stages of their career and less experienced.

According to Hanson and Richards (2019), this unpreparedness causes many researchers to write embodied experiences out of qualitative research, with potentially negative consequences for the production of knowledge. While ignoring such surplus data can be a coping mechanism, "excluding these observations also prevents scholars from consciously integrating what these experiences tell us about the field" (Hanson and Richards 2019, 165). Instead of seeing bodies as barriers, then, they advocate for an embodied approach that recognizes that "all data and knowledge emerge from experiences, conversations, and interactions shaped by the bodies that engage in them" (Hanson and Richards 2019, 16). Such an approach invites scholars to rethink the very notion of the "awkward surplus" by reintegrating it into their research and to consider the ways it may affect what is considered data. In doing so, it is part of a growing literature that calls for the recognition of "the messy reality of fieldwork" (Bliesemann de Guevara and Bøås 2020, 1) and the reconceptualization of failure as "productive rupture" (Bliesemann de Guevara and Kurowska 2020) and a "potential site for learning" (Cole 2020, 88).

As my own reflections show, such a perspective can be deeply empowering. Once I paid attention to my embodied experience and how it related to power dynamics in the field, I understood that rather than preventing me from collecting data, this experience was, in and of itself, *highly valuable data.* First, it offered important insights into local gender roles and stereotypes. Not only did it reveal the distinction between local and foreign women, and the limited accountability of men in relation to women who are not their wives, but also women more generally, as became evident in Rustam's refusal to discuss the issue with Zhanna despite her being Abkhaz. Second, it foregrounded the vulnerable position of those locals who have any connection with Georgia – be it through NGO and civil society work, or family relations. As my experience showed, any such association, regardless of its intensity and meaning, can be used as a weapon against a person for reasons not necessarily political.

But most importantly, my embodied experience exposed the pervasiveness of mistrust. While being betrayed by a participant instilled a certain paranoia, which, in turn, had an impact on my ability and willingness to connect to other people and build the kind of intimate connections with strangers that tend to be expected of ethnographers, this was in many ways a common behavior among people living in Abkhazia. As such, it provided profound insight into the social dynamics of the place I was studying. One could even say that Rustam's betrayal and the events that unfolded were my "Balinese cockfight," to reference Clifford Geertz's (2005, 56–59) famous rapport tale, in which he and his wife were unexpectedly accepted by the village community after running away from the police during an illegal cockfight. While my experience first engendered a profound sense of failure in terms of becoming an insider, making sense of what happened in collaboration with locals who had gone through similar experiences created long-lasting bonds and, ultimately, brought me closer to an insider position. To this end, my exposure to betrayal had an invaluable influence on the direction of my research. Most importantly, it steered my interest away from the predominant focus on interethnic tensions to intraethnic divisions and socioeconomic conflict, thus heightening my awareness of the impact of war, displacement, and nonrecognition on social relations more generally. This accordingly enabled me to question the widespread assumption that relations among co-ethnics are automatically relations of trust.[17]

Concluding Reflections

This chapter explored the double burden of conducting fieldwork in Abkhazia as a young, Western female researcher. Coming from a country that does not recognize Abkhazia's statehood, I was confronted with hostility and political pressure and was also concerned about potentially causing reputational damage to those assisting me. While working independently appeared to be an easy solution at first, it was complicated by the fact that I was a foreign woman with no husband (or other male protector) by her side. Consequently, I not only needed to navigate potentially hostile encounters but also had to worry about being taken advantage of by male contacts.[18] It was after my interactions with Rustam and his betrayal that I became so paranoid and distrustful that, at one point, it seemed difficult to continue my work. However, after an initial period of despair and extreme self-doubt about my abilities as a researcher, and particularly as an ethnographer, I recognized the unique value of this experience, which provided in-depth insights into Abkhazian society.

The chapter hence engaged in a more honest discussion of some of the challenges that I faced during my fieldwork. In doing so, it provides a fresh, embodied perspective that allows us to rethink these challenges not simply as limitations to an idealized version of fieldwork based on "white men's experience" (Hanson and Richards 2019, 40), but as essential data that can tell us much about what is happening in the locations we study. However, highlighting the value of a more embodied approach should not be misread as encouragement to actively seek out difficult experiences. Rather than seeing challenges during fieldwork as part of a necessary and inevitable "rite of passage," it encourages a frank discussion about the reactions our bodies can elicit in the field, and how this impacts us as researchers. Moreover, instead of placing the burden on researchers as individuals or on the people we encounter in the field, it situates them within the wider power relations both in the societies that we study as well as our own. While the chapter suggests that a reflective, embodied approach to difficult situations can enrich our research, and hence knowledge production more broadly, this is by no means intended to put pressure on already vulnerable young scholars to tolerate risky situations for the sake of getting "good" data.

Against this background, I want to conclude this chapter with a brief reflection on some of the practical changes that I implemented in the light of my experience as well as the things that I wish I had done differently. While I found it extremely helpful to reconceptualize certain challenges as windows into the society that I studied and to think of mistakes "as gifts" (Fujii 2017, 48), they nevertheless caused me to modify my own expectations of what fieldwork should be like and the practices that it involves. Most importantly, acknowledging the emotional toll that fieldwork took, I began to embrace the shorter, two-month research stays that I was forced to undertake due to visa issues. Having breaks between research trips allowed me to more critically reflect on the data that I had already collected and gave me a better sense of what still needed to be done. In fact, I noticed that the more time I spent in the field, the more difficult it became to focus intellectually. Over time, I became increasingly aware of the temporal dynamics of fieldwork (which, of course, vary from person to person). Usually, I was most proactive, curious, and open-minded in the first few weeks of my stay. After that initial period, I tended to narrow down my network and spend more time with specific people, which provided more intimate insights but also meant that I became increasingly immersed in certain dynamics and the, at times, tense local atmosphere. Interestingly, I once again noticed a similar cycle among some of my interlocutors, who frequently went through phases when they stressed how much they "needed a break" from the people and activities in their daily lives.

Taking breaks not only allowed me to recuperate but also ensured that my network remained as diverse as possible. Every reentry was an opportunity to meet new people and focus on new angles. From an ethical perspective, this implied that I became less dependent on specific contacts, which, in turn, made me feel less vulnerable. Over time, I learned that there are always going to be people willing to help and that there was no need to force myself to stay in touch with someone I was not comfortable with just for the sake of collecting data. This knowledge also enabled me to be more selective and resist opportunities that seemed tempting yet risky. As the pressure to collect data decreased over time, conducting fieldwork naturally became less stressful, making it easier to avoid certain dilemmas.

Once again, this does not mean that one should simply persevere in the early stages of fieldwork. In hindsight, I wish I had allowed for a longer exploratory phase. Such a period not only eases the pressure to get as much information as possible from the beginning but also gives one a good sense of the possible limitations and obstacles of conducting research in a field site, allowing time to adapt one's expectations and approach early on. In some cases, it might even lead researchers to opt for a more limited form of immersion as the most ethical choice given their circumstances. As Jana Krause noted:

> [i]mmersion is not simply the product of a learned skillset the researcher wields independently of their embodied identity. Obstacles that result from this identity, such as gendered and racial bias, cannot and perhaps should not always be overcome by sheer persistence. Instead, researchers may choose uneven immersion to navigate a field site ethically, weighing the benefits of immersion against its challenges, such as the performance of a specific notion of masculinity or femininity for maintaining access.
>
> (Krause 2021, 338)

My fieldwork experience made me recognize the importance of adopting a "flexible practice" which "avoids the mistakes made by others, while paying attention to the volatility, context-specificity and long-term and wider effects of research in violent or illiberal contexts" (Bliesemann de Guevara and Bøås 2020, 17). Such a perspective has empowered me to engage in more honest – and productive – conversations with my supervisors and colleagues who work in similar areas and to build a support network that will be essential for future projects to come.

Notes

1 This chapter draws on research funded by the UK's Economic and Social Research Council. I am grateful to the editors for their helpful comments on earlier versions of this chapter. My special thanks go to the people in Abkhazia

who shared their insights with me and helped me navigate a complex field site. I would also like to express my gratitude to my supervisors at University College London, Richard Mole and Jan Kubik, for their unconditional support before, during, and after my fieldwork.

2 For an exception, see Shesterinina (2019).

3 However, it is illegal under Georgian law to enter Abkhazia via Russia (see GOV.UK 2022).

4 As noted in a report by the International Crisis Group, "[c]ontrary to widespread misconceptions, there was relatively little combat in and around Abkhazia" (2010, 1) during the 2008 war between Russia and Georgia. The region mainly functioned as a transit route for Russian troops, and the fighting that did take place was limited to the mountainous Kodori Gorge area in the far northeast of Abkhazia.

5 The linguistic juxtaposition of Abkhazia and Georgia is made for practical reasons and does not imply any judgment regarding Abkhazia's legal status.

6 Place names are disputed between the two conflicting parties. Where there are different toponyms, I chose the ones used by my interlocutors.

7 To some extent, being a woman could mitigate the exposure to hostility, as it was usually considered less appropriate to exhibit outright threatening or hostile attitudes toward a woman.

8 The most famous example is Bronislaw Malinovski's (1967) posthumously published fieldwork diary.

9 All names used in this chapter are pseudonyms.

10 Women are often advised to wear "fake wedding rings" (Johnson 2009, 323) to limit unwanted male attention. This, however, can raise ethical questions around self-representation and authenticity. I personally found it more stressful to lie in the contexts where the people I met could become key interlocutors.

11 While I agree with Helene Thibault (2021) that marriage proposals can yield important ethnographic data, it is nevertheless important to acknowledge the stress and discomfort that they can cause for young, female researchers.

12 This also reveals a stark difference in how material interests and personal relationships were seen to be linked. In my understanding, friendship and money – or what Beek calls "the emotional and instrumental dimension of relationships" (2018, 64) – had to be clearly separated. However, as I began to understand, in Abkhazia, it is common for people to rely on friends and family for money.

13 While I did acquire some basic Abkhaz language skills, it was not enough to follow a conversation.

14 As Fujii points out, "no researcher is a 'true' insider or outsider from beginning to end. Many will occupy both categories at various points in time or shift from one to the other" (2017, 19; see also Herod 1999).

15 Abkhazia's postwar population continued to be multiethnic. According to the 2003 census, ethnic Abkhazians constitute only 44 percent of the overall population. Other ethnic groups include Armenians (21 percent), Russians (11 percent), and Georgians (20.6 percent) (Trier, Lohm and Szakonyi 2010).

16 For a recent discussion of the differences between participant observation and observant participation, see Seim (2021).

17 For more details, see Peinhopf (2020, 2021).

18 It is important to note that while I focus on men taking advantage of women in this chapter, betrayal is by no means exclusively a male phenomenon.

References

Beek, J. 2018. "How Not to Fall in Love: Mistrust in Online Romance Scams." In *Mistrust: Ethnographic Approximations*, edited by F. Mühlfried, 49–69. Bielefeld: Transcript. DOI: 10.14361/9783839439234-003.

Blakkisrud, H. and P. Kolstø. 2012. "Dynamics of *De Facto* Statehood: The South Caucasian *De Facto* States between Secession and Sovereignty." *Southeast European and Black Sea Studies* 12 (2): 281–298. DOI: 10.1080/14683857.2012.686013.

Bliesemann de Guevara, B. and M. Bøås. 2020. "Doing Fieldwork in Areas of International Intervention into Violent and Closed Contexts." In *Doing Fieldwork in Areas of International Intervention: A Guide to Research in Violent and Closed Contexts*, edited by B. Bliesemann de Guevara and M. Bøås, 1–20. Bristol: Bristol University Press.

Bliesemann de Guevara, B. and X. Kurowska. 2020. "Building on Ruins or Patching Up the Possible? Reinscribing Fieldwork Failure in IR as a Productive Rupture." In *Fieldwork as Failure: Living and Knowing in the Field of International Relations*, edited by K. Kušić and J. Záhora, 163–174. Bristol: E-International Relations Publishing.

Bryant, R. 2014. "Living with Liminality: *De Facto* States on the Threshold of the Global." *Brown Journal of World Affairs* 20 (2): 125–143.

Caspersen, N. 2012. *Unrecognized States: The Struggle for Sovereignty in the Modern International System*. Cambridge: Polity Press.

Chamberlain-Creanga, R. 2008. "The 'Transnistrian People': Citizenship and Imaginings of 'The State' in an Unrecognised Country." In *Weak State, Uncertain Citizenship: Moldova*, edited by M. Heintz, 103–124. Frankfurt am Main: Peter Lang.

Clogg, R. 2008. "The Politics of Identity in Post-Soviet Abkhazia: Managing Diversity and Unresolved Conflict." *Nationalities Papers* 36 (2): 305–329. DOI: 10.1080/00905990801934371.

Cole, L. 2020. "Tears and Laughter: Affective Failure and Mis/Recognition in Feminist IR Research." In *Fieldwork as Failure: Living and Knowing in the Field of International Relations*, edited by K. Kušić and J. Záhora, 76–90. Bristol: E-International Relations Publishing.

Costello, M. 2015. *Law as Adjunct to Custom? Abkhaz Custom and Law in Today's State-Building and 'Modernisation' (Studied through Dispute Resolution)*. PhD dissertation, University of Kent.

Fujii, L. 2017. *Interviewing in Social Science Research: A Relational Approach*. New York: Routledge.

Fujimura, J. 2006. "Sex Genes: A Critical Sociomaterial Approach to the Politics and Molecular Genetics of Sex Determination." *Signs: Journal of Women in Culture and Society* 32 (1): 49–82. DOI: 10.1086/505612.

Gallaher, C. 2009. "Researching Repellent Groups: Some Methodological Considerations on How to Represent Militants, Radicals, and Other Belligerents." In *Surviving Field Research: Working in Violent and Difficult Situations*, edited by C.L. Sriram et al., 127–146. London: Routledge.

Geertz, C. 2005. "Deep Play: Notes on the Balinese Cockfight." *Daedalus* 134 (4): 56–86.

GOV.UK. 2022. "Travel Advice Georgia." Available at: https://www.gov.uk/foreign-travel-advice/georgia (accessed on February 22, 2022).

Hanson, R. and P. Richards. 2019. *Harassed: Gender, Bodies and Ethnographic Research*. Oakland: University of California Press.

Herod, A. 1999. "Reflections on Interviewing Foreign Elites: Praxis, Positionality, Validity, and the Cult of the Insider." *Geoforum* 30 (4): 313–327. DOI: 10.1016/S0016-7185(99)00024-X.

Human Rights Watch. 1995. *Georgia/Abkhazia: Violations of the Laws of War and Russia's Role in the Conflict*. Helsinki: Human Rights Watch Arms Project/Human Rights Watch.

International Crisis Group. 2010. "Abkhazia: Deepening Dependence." Europe Report No. 202.

Johnson, J. 2009. "Unwilling Participant Observation among Russian *Siloviki* and the Good-Enough Field Researcher." *PS: Political Science & Politics* 42 (2): 321–324. DOI: 10.1017/S1049096509090647.

Kabachnik, P. 2012. "Wounds that Won't Heal: Cartographic Anxieties and the Quest for Territorial Integrity in Georgia." *Central Asian Survey* 31 (1): 45–60. DOI: 10.1080/02634937.2012.647842.

Kolstø, P. and H. Blakkisrud. 2013. "Yielding to the Sons of the Soil: Abkhazian Democracy and the Marginalization of the Armenian Vote." *Ethnic and Racial Studies* 36 (12): 2075–2095. DOI: 10.1080/01419870.2012.675079.

Krause, J. 2021. "The Ethics of Ethnographic Methods in Conflict Zones." *Journal of Peace Research* 58 (3): 329–341. DOI: 10.1177/0022343320971021.

Lynch, D. 2007. "De Facto 'States' around the Black Sea: The Importance of Fear." *Southeastern European and Black Sea Studies* 7 (3): 483–496. DOI: 10.1080/14683850701566484.

Malinovski, B. 1967. *A Diary in the Strict Sense of the Term*. London: The Athlone Press.

Ministry of Foreign Affairs of the Republic of Abkhazia. 2022. "Consular Service." Available at: http://mfaapsny.org/en/consular-service/ (accessed on February 22, 2022).

Mühlfried, F. 2019. *Mistrust: A Global Perspective*. Cham: Palgrave Macmillan. DOI: 10.1007/978-3-030-11470-1.

Ó Beacháin, D. 2012. "The Dynamics of Electoral Politics in Abkhazia." *Communist and Post-Communist Studies* 45 (1–2): 165–174. DOI: 10.1016/j.postcomstud.2012.03.008.

Peinhopf, A. 2020. *Conflict and Co-Existence: War, Displacement and the Changing Dynamics of Inter- and Intra-Ethnic Relations in Abkhazia*. PhD dissertation, University College London.

Peinhopf, A. 2021. "The Curse of Displacement: Local Narratives of Forced Expulsion and the Appropriation of Abandoned Property in Abkhazia." *Nationalities Papers* 49 (4): 710–727. DOI: 10.1017/nps.2020.30.

Seim, J. 2021. "Participant Observation, Observant Participation, and Hybrid Ethnography." *Sociological Methods & Research*: 1–32. DOI: 10.1177/0049124120986209.

Shahnazarian, N. and U. Ziemer. 2014. "Emotions, Loss and Change: Armenian Women and Post-Socialist Transformations in Nagorny Karabakh." *Caucasus Survey* 2 (1–2): 27–40. DOI: 10.1080/23761199.2014.11417298.

Shesterinina, A. 2014. *Mobilization in Civil War: Latent Norms, Social Relations, and Inter-Group Violence in Abkhazia*. PhD dissertation, York University.

Shesterinina, A. 2019. "Ethics, Empathy, and Fear in Research on Violent Conflict." *Journal of Peace Research* 56 (2): 190–202. DOI: 10.1177/0022343318783246.

Thibault, H. 2021. "'Are You Married?': Gender and Faith in Political Ethnographic Research." *Journal of Contemporary Ethnography* 50 (3): 395–416. DOI: 10.1177/0891241620986852.

Trier, T., H. Lohm and D. Szakonyi. 2010. *Under Siege: Inter-Ethnic Relations in Abkhazia*. London: Hurst.

Wood, E. 2006. "The Ethical Challenges of Field Research in Conflict Zones." *Qualitative Sociology* 29 (3): 373–386. DOI: 10.1007/s11133-006-9027-8.

2 Doing Fieldwork (Not Quite) at Home

Reflecting on an Expat's Positionality in Lithuania

Rasa Kamarauskaitė
UCL SSEES

Introduction: Positionality as Speaking from a Specific Location

The positivistic approach to fieldwork asserts that researchers should set aside their beliefs, values, and personal circumstances to ensure an impartial research outcome. This approach has been criticized by social scientists (see, for example, Fook 1999 and Gardner 1999 among many others). The critics of the positivistic approach argue that researchers' subjectivity needs to be acknowledged and recognized as it forms an integral part of the research process. For example, Gardner (1999) argues that "everyone writes from the specific locations" (Gardner 1999, 51), and therefore it is impossible for researchers to occupy a neutral position or to produce solely neutral results in their research. To produce convincing and transparent research findings, the researchers' beliefs, values, and circumstances must therefore be acknowledged, made explicit, and incorporated into the research process. In this context, we speak of reflexivity to refer to disclosure of the researcher's beliefs, values, subjectivities, and any other circumstances that might influence the research methodology and the more extensive production of knowledge (Fook 1999; Hellawell 2006). While reflexivity can be understood and defined in many ways, endogenous and referential reflexivity in particular can help us better understand researchers' positionalities. Endogenous reflexivity (also called subjectivity) helps researchers understand how their self is formed and positioned, whereas referential reflexivity focuses on how one is viewed by others (e.g., why some actions and positions are encouraged or restrained in a certain environment, and how the environment is shaped by the researchers' actions) (Noh 2019). Throughout this chapter, I will frame my experiences of conducting fieldwork "at home" through the lens of both endogenous and referential reflexivity.

In order to truly be reflexive, though, one must equally understand one's own positionality. Positionality premises that an individual's multiple identities inform their relation to and with others (Noh 2019). It is therefore only through recognizing their "position" vis-à-vis other people,

DOI: 10.4324/9781003144168-4

that researchers are able to be reflexive. In this chapter, I show how my identities as a Lithuanian, expat, and nonheterosexual female positioned me as an insider, an outsider, and an inbetweener (Milligan 2016). I will discuss how the multiple positions resulted in inclusions and exclusions (perceived by me and my research participants) during my data collection and impacted my ability to recruit participants, build trust, and establish rapport. I argue that I was mostly perceived by the research participants as an insider, but due to my "homecomer" (Schuetz 1945, 369) position, I saw myself as an outsider and as an inbetweener (Milligan 2016, 248). I therefore discuss how these positionalities helped me, yet, paradoxically, made it more difficult for me to navigate the complexities of my fieldwork. The chapter concludes by emphasizing that the more intersections there are between researchers and their research participants, the more complex the relationships become.

Conducting Fieldwork One Bit at a Time

In this section, I briefly discuss the rationale behind conducting my PhD fieldwork in Lithuania over a series of short periods. Lithuania is a Baltic state that formerly belonged to the Soviet Union. Following the collapse of the USSR in 1991, ethno-religious nationalism (Sen 2020), i.e., national belonging based on a conflation of religious and ethnic attributes, became the main ideology that defined national identity in Lithuania. Ethno-religious nationalism proclaimed the heterosexual nuclear family to be the only legitimate social unit and, consequently, delegitimized other forms of cohabitation and sexual expression, including homosexuality. As a consequence, according to ILGA-Europe (2022)[1] (the Annual Review January to December 2021), Lithuania is one of the most homophobic societies in Europe. The ILGA-Europe Annual Review maintains that two thirds of Lithuanian society still oppose same-sex partnerships.

Because of this, Lithuanian nonheterosexual people, unsurprisingly, encounter significant difficulties in many aspects of their lives, including visibility. It was for these reasons that I chose to center my doctoral studies on researching the multiple (in)visibilities that Lithuanian nonheterosexual people face in their everyday lives, as well as exploring how they make decisions in terms of revealing and/or disclosing their sexualities. To do so, I collected descriptive data, recounting the lived experiences, values, and decisions that contributed to their decisions around (in)visibility through semi-structured in-depth interviewing and observation. Before pursuing my thesis, I only personally knew three research participants. The rest were recruited by snowballing and by a participant call sent out via various online LGBTQ (lesbian, gay, bisexual, transgender, and queer) groups.

In contrast to a common understanding of fieldwork, where an ethnographer goes to the research location and stays there for a significant amount of time (Evans-Pritchard 2004[1951]), I decided to split my presence in the physical sites into small segments: 15–20 approximately 7–10-day visits to

Lithuania dispersed over an 18–24-month period. My decision was made by my use of grounded theory[2] as a methodology to analyze my data, which required examining it from the very early stages of research, and, thus, could best be achieved through short pauses in the data collection. Since data analysis requires a certain degree of disconnection and distance from the field (Powdermaker 1967), I chose to analyze my data in a university environment in the UK. In addition, as a self-funded student with work and family commitments in the UK, I was only able to leave the country for fieldwork for short periods. Thus, my fieldwork design was informed not only by objective reasoning but also partially by my marital and economic status. From my experience, knowledge production, the field, and home should be understood as situated in a broader context of personal and professional institutional relations (Till 2011).

Notably, Evans-Pritchard argues that ethnography is only possible through prolonged stays in the field (Evans-Pritchard 2004[1951]). However, these views assume that a field is a geographically bounded, physical location and that data collection and data analysis are two distinct but consecutive stages of the research process. This is not entirely the case. Shore (1999, 26) criticized this position as "increasingly outdated and untenable" because the personal ties and cultural exchanges created in "the field" continue even after data collection is officially complete and, in various ways, inform the data analysis during the process of writing up. Staying in touch with my research participants after conducting interviews has been a regular practice in my PhD fieldwork. Interviewees would usually initiate this contact. They would "befriend" me on social media platforms or contact me via email. Our contacts ranged from casual comments on social media content to a steady exchange of emails and regular meetings when I visited Lithuania. Keeping in touch and sharing details about our lives added another layer of understanding and interpretation for the data that I have collected through interviews and observations. For example, a research participant's shared photos of their summer cottage prompted me to reread the interview passages related to the cottage. As a result, it advanced my thinking about the meaning of a safe space.

Still, regularly moving between the two countries helped me to also understand and reflect on my positionality and see firsthand how my "position informs the ways in which [I] approach and experience the field" (Gardner 1999, 51). In particular, my positionality as a nonheterosexual Lithuanian woman living abroad researching nonheterosexual communities in Lithuania highlighted several issues regarding the relationship between the home, the field, and my role in both, which I had not foreseen prior to my data collection. I outline these in the next sections by explaining how (self-)positioning myself within the field influenced my PhD fieldwork progress. I also reflect on my fieldwork preparation and presume what actions and decisions may have helped me to better prepare for my field research.

Deceptive Belongings: Being a "Homecomer"

Traditional accounts of ethnographic research focus on exploring and immersing oneself in a culture that is not one's own. In this way, the researcher's position as a stranger is considered ideal for ethnographic research, and their goal is to understand a society that is different from what they already know (Powdermaker 1967; Evans-Pritchard 2004[1951]). They often do this by traveling to another location where they have had no prior involvement. This does not have to be a foreign country but could equally be a new environment within the same country or even locale. The "field" can therefore be defined as a site of alienation and investigation, while "home," by contrast, is a site of belonging and familiarity (Knowles 1999). However, for researchers who engage in fieldwork within familiar contexts, such as at home or near their own places of residence, the relationship between the home and the field requires some further renegotiation as "a presumption that 'home' is stationary while the field is a journey away" (Amit 2000, 8) does not reflect their positionality. To scholars who return to countries of considerable familiarity to conduct research, such as expats returning to their countries of birth, the relationship between home and the field is even more complex. If we define "home" as endowed with the sense of belonging, these individuals often develop multiple sites of belonging, and thus feelings of being an insider or an outsider often coexist, rather than standing in opposition to one another (Zhao 2017).

The sense of simultaneously feeling like an outsider and insider is often considered as taking a position, or being positioned as, an "inbetweener" (Milligan 2016, 248) and/or a "homecomer" (Schuetz cited in McNess, Lore and Crossley 2013, 303). While the former refers to a researcher who is "neither entirely inside or outside" (Milligan 2016, 235) due to a wide array of overlapping or diverging identities, such as those associated with class, race, or profession, the latter specifically refers to someone who simultaneously feels as an outsider and insider due to the time spent apart from a once-familiar environment (Schuetz 1945, 369). Unlike the inbetweener or an outsider, the homecomer is a person who expects a sense of familiarity and comfort within an environment that they once knew (Schuetz 1945). Yet upon their return, the homecomer needs to come to terms with the fact that the environment is no longer as familiar as it was, hence giving rise to feelings of uncertainty, ambiguity, and of being in-between. Although these concepts are all very closely linked, I feel that my experience resembles that of a homecomer, as I grew up in Lithuania and returned to conduct my research after spending a decade in the UK. Being an expat, to me, means simultaneously belonging to two places and, yet, not fully belonging to either. Indeed, throughout my life as a Lithuanian expat, I have developed two distinctive attachments to the UK (specifically London) and to Lithuania and, therefore, call both places "home." Nonetheless, I am constantly reminded that I do not fully belong in either place because I cannot provide a simple answer to the question, "where are you from?" As I spent more time in London, I came to observe that my sense of belonging to Lithuania was mostly rooted in a nostalgic longing for the past and not to the contemporary version of the country – it was a feeling that had a quality of "a mirage which disappear[ed] in the act of travelling to it" (Knowles 1999, 64). I quickly notice how defamiliarized I am with Lithuania (Munthali 2001): I did not recognize the famous people in the magazines, the renovated streets, and, after visiting all of my close friends and family members, I realized that I had nowhere to go and nothing to do. In other words, I believe that my feeling in-between stems from a disrupted feeling of entirely belonging to one particular context.

At the same time, going back to Lithuania to do fieldwork still provided me with an opportunity to reconnect with my former life, as well as giving me an option to temporarily escape from my current one in London. Even though I always felt a dual sense of belonging, the constant moving back and forth between the two places made the feelings more acute, and I had an opportunity to reflect on those feelings during my fieldwork. In many ways, my fieldwork not only took me away from one of my homes but also gave me a chance to return to another one. I experienced this in many different settings. For instance, I remember casually chatting with one of my informants, with whom I had a good rapport, when the conversation took a slightly nostalgic turn as she asked: "[o]f all the countries in the world, why the heck did you choose to do this research here?" Being inspired by our soul-searching conversation, I gave her the most honest answer: "[h]aving been away for so long, I think I am kind of trying to find my way back." It was not the first time that I was asked the same question in either Lithuania or in London, but all the other times, I pointed out the importance of recording and making visible the experiences of Lithuanian nonheterosexual people and the lack of knowledge surrounding these issues. The answers I gave during these other conversations were also true and honest, but when taken together, reveal the multiple and complex motivations behind individuals' research projects, particularly in terms of the researchers' backgrounds.

As is evident, the sense of being in-between (Zhao 2017) was characteristic to my self-positioning in the field and my fieldwork experiences more largely. On the one hand, I both saw myself and was perceived by the research participants as a Lithuanian woman: I encountered no language barriers while in the field and was able to share similar experiences with my research participants about growing up in Lithuania, including memories from school and university, in addition to common knowledge about life in the country. To many participants, these shared common experiences indicated that I was indeed "one of them," and, as a result, some of the research participants became suspicious when I started to ask them questions about the issues which seemed too common sense to query. The clearest example is from an interview with a middle-aged gay man who was very surprised to hear some of my questions. To him, the answers seemed quite obvious, and consequently, he remarked: "but you *have* lived here. Don't you know how things normally go?" This engagement proved difficult for me and my research because my informants perceived *me* as a native and therefore "all knowing" about common-sense sociocultural matters (Munthali 2001). In order to resolve these challenges, I therefore often felt the need to explain to my research participants that I left Lithuania quite a long time ago and that I have forgotten, or have not personally experienced, many of the things that they might see as self-explanatory. This approach indeed helped to encourage my interlocutors to be more vocal and descriptive when explaining certain aspects of their lives and also allowed me to reposition myself more accurately in the field as a researcher and as someone who is "not quite" from there.

Deceptive Belongings: An Interloper

Still, my homecomer positionality as a returning expat significantly informed my interactions with my research participants. In particular, my socioeconomic and class position shaped my rapport in the field. For the purposes of this chapter, I define class as a social position in society primarily informed by economic/material attributes, which serves as a basis for social and symbolic characteristics (Bukodi and Goldthorpe 2018). When I reflect on my own class position, I consider myself as a socioeconomically vulnerable inbetweener, as my position in a labor market and pattern of transient residence prevents me from clearly self-identifying with a particular social class and informs my subjective positionality as an interloper (Ryan and Sackrey, cited in Johansson and Jones 2019, 1528). Even though I always saw my employment as a means of pursuing higher education, spending more than a decade in working-class occupations has prevented me from trivializing this experience as merely a passing stage in my life until I am able to get a "real" (i.e., middle-class) academic or similar job. At the same time, my academic, middle-class employment is not yet established enough to provide me with either a material, social, or symbolic base (Bradley cited in Cederberg 2017) that would allow me to lucidly self-identify as middle class. As I occupy both spheres "with a marginal sense of membership" (Ryan and Sackrey, cited in Johansson and Jones 2019, 1528), I find myself an interloper regarding my socioeconomic/class position.

Sultana (2007) argues that homecoming Western-educated scholars need to critically evaluate their positionality from class and educational perspectives, as different class and educational statuses might result in unequal relationships developing in the field. For instance, a sense of alienation and/or mistrust between the researcher and research participants may arise and even impede the building of rapport and the process of data collection, since people in more vulnerable social positions often react deferentially to urban, educated elites (Sultana 2007). Moreover, the perceived superiority of the researcher might also create issues regarding informed and independent consent (Katyal and King 2011), a prerequisite for conducting ethical field research. However, during my fieldwork, I did not experience the power dynamics as described above. Rather than as an admirable, awe-inspiring scholar from the privileged West, I was actually viewed as a struggling student who needed help. This attitude was informed, I assume, by the relatively well-established and stable social positions (working class as well as middle class) of my research participants, in contrast to my somewhat unstable socioeconomic position as a student. My research participants' perception of my social standing was furthermore reinforced by their own intimate knowledge of the realities of economic migration to Western countries, like the UK, gained from, say, friends and relatives, or even through their own experiences. As one of my participants stated when describing her former life in London: "crowds on the tube, constant rush, and low quality of life."[3] My sharing of details about my life in the UK reflected their knowledge about migrant life abroad, and therefore, I was not necessarily perceived as a professional, middle-class expat who was building a successful career abroad (Saar and Saar 2019). Despite my education, my research participants instead bracketed me as a struggling East European migrant who was also a student and not vice versa. In other words: my position was perceived based on my role in the labor market and patterns of residence (Bukodi and Goldthorpe 2018) and only secondarily on my educational experience.

Notably, being perceived as socioeconomically vulnerable, in addition to my national and sexual identities, helped me to build rapport with my interviewees and advance my data collection. In particular, my work and study experience in the UK helped me to bond with research participants who had worked and/or studied abroad and with those who had similar occupations in Lithuania. The more affluent research participants also often felt inclined to help a "struggling student" by offering to assist me in recruiting new participants or by insisting to pay for my share of café and restaurant bills during my fieldwork. These few examples again reinforce how my positionality impacted on my relationships in the field.

Deceptive Belongings: Being Nonheterosexual

The particularities of belonging within Lithuanian society as a nonheterosexual female also shifted my positionality from insider to outsider during the course of my fieldwork. From the commencement of my research, I felt very committed to my role as a representative of nonheterosexual Lithuanian's experiences – I believed that I was representing my people. Similar to the idea of universal womanhood proposed by the first wave feminists (for more, see Gardner 1999), I had a somewhat unarticulated conviction that a shared sexual category represents similar experiences which lead to personal bonds. My belief that I was "one of them," or an insider, helped me to maintain focus and dedication during my data collection. I was eager to get back into the field to collect more data and was able to stay active and engaged for long hours while in the field – sometimes conducting up to three in-depth interviews per day, typing up field notes in the evening, and setting off to observe a party or an event that same night. The idea that I was doing something important for my people was a driving force, which helped me to stay focused and engaged in the field.

However, as my fieldwork progressed, I began to realize the problems that can arise when researching a topic with which I felt personally connected. The idea that my research is a tool to represent people similar to me made it difficult to maintain the necessary cognitive and emotional detachment from the field. The distinction between my own self and others (as my research participants) was therefore diminished, and hence, I felt that by representing them, I was to a certain extent representing myself. Notably, the "Othering" of research participants and the reinforcement of distance between the researcher and their participants as a practice has often been criticized as contributing to an unequal relationship in the field that empowers researchers while subjugating research participants (Ellis 1995). However, in my case, the conviction that I was representing my own people made it difficult for me to fully accept and evaluate the data that did not align with my own self-image as a nonheterosexual woman, as someone who is in a monogamous relationship with another woman. I noticed that I was not able to approach my observations critically and objectively as my own emotions and opinions toward my data interfered with my critical thinking and I found it difficult to remove myself from my own biases. At that time, I was in Lithuania, and my original plan was to conduct intensive fieldwork. However, I decided to postpone my meetings for two days, to stay by myself, and to reflect on my experiences and feelings. Once I realized which interactions in the field were most frustrating for me, I started to read through my fieldwork diary, where I was writing down my impressions and ideas. Unlike the field notes, which were more focused on research-related descriptions, the fieldwork diary was a more personal account focused on my reactions, mood, and thoughts that arose in the field. Reflecting and reading helped me to understand that the source of my difficulties in approaching my data critically was my conviction that the research participants and I are "the same" and their life choices and values also represent mine and vice versa. The realization has prompted me to reevaluate my sexual identity, my relationships with the research participants and role in the field. I asked myself: What am I doing here? What is my goal here? My answer was "to represent the data as honestly and objectively as I possibly can." In that moment I realized that I could only achieve it by repositioning myself as "the other" in relation to my research participants. Therefore, I did not create a boundary between myself and my respondents to exercise power but rather to engage with the data and to adequately represent it (Owton and Allen-Collinson 2014). Had I not decided to step away from the field, and to examine my bias, I believe my data analysis might have turned out less critical and objective. Exposing and critically evaluating my biases, I believe, helps to ensure the transparency of the research process (Given 2008). As Powdermaker aptly maintained:

simultaneous involvement and detachment "is the heart of the participant observation method" (1967, 9). Involvement is necessary for gathering data and detachment for analyzing it critically.

As my fieldwork progressed, I eventually realized that belonging to the same group as my research participants was actually based on a *presumed* sense of commonality (Owton and Allen-Collinson 2014). Having left Lithuania in my early 20s, I had no connection to the Lithuanian LGBT+ community, nor had I experienced any of the events that were characteristic of their lives, as described to me during my field research. For example, I was not forced to hide my sexuality at work, think twice before taking my wife's hand in public, nor was I prohibited by law from getting married. Nevertheless, because of my national and, most importantly, my sexual identity, I was unquestionably "one of them" in the eyes of my research participants and was immediately accepted as an insider by every one of them. Accordingly, I did not go through the same period of transition, when the ethnographer needs to gain the trust of the group studied (Powdermaker 1967), nor did I arouse any suspicion or distrust on the part of my research participants. Rather, I was warmly welcomed to their homes, driven around, taken out for drinks, and offered help with finding more participants. In fact, one evening, I shared my surprise about the openness of the people to which one of my research participants replied: "But of course, Rasa. You are one of us." In contrast to how I perceived myself, it appeared obvious to her that people would be open and welcoming, as they were seeing me first as a lesbian Lithuanian and second as a researcher (Zhao 2017). Hence, even though I did not perceive myself as a true local, I was nonetheless perceived by others as such.

However, due to the intersection of identity categories between myself and my research participants, and the fact that they often downplayed my role as a researcher and rather saw me as a fellow lesbian Lithuanian, some queries around ethics arose in the conduct of my research, specifically in terms of obtaining informed consent. I quickly realized that even though I had clearly informed them about the goal of my presence in Lithuania, certain conversations seemed to be intended to remain private and off the record. Furthermore, as my research progressed, my relationships with several key informants grew into more personal connections over time, which elicited some ethical questions about my dual role of researcher and friend in the field. Specifically, it was when some of the research participants became friends or acquaintances that I became more concerned about sustaining the relationship, as well as more emotionally involved and reflexive about our interactions (Owton and Allen-Collinson 2014). As our relationships strengthened and moved from participant to friend, I also realized my duty of care for them and that I did not want to use our friendship as "a guise strategically aimed at gaining further access" (Tillmann-Healy 2003, 739). In other words, I found it unethical to use the newly developed personal relationships in the field to collect more data.

In order to ensure my research aligned with my ethical responsibilities to my participants, and to avoid potential repercussions about using certain information intended to be private (Ellis 1995), I therefore began to compartmentalize my roles in the field and to distinguish between my roles as friend and as researcher. In some instances, my position did change within the same interaction. For example, I once met up for lunch with an informant who had become a friend, to offer her some support after a particularly painful breakup. As our conversation developed, she offered me significant insights into her relationship that I found useful for my research. After we finished lunch, I carefully asked her if it was okay for me to reflect on her story for my research and assured her that it is absolutely fine if she did not feel like it. It was only after getting her permission that I went home and wrote up our conversation in my field notes. Importantly, I did not approach all situations in this way. In other situations, in which my informal conversations with informants could have served as brilliant opportunities for data collection, the circumstances precluded me from doing so, as they were either too painful or too personal and I believed it was inappropriate to even ask about the possibility of adding it to my research. In this way, approaching each situation on a case-by-case basis, considering the larger ethical questions, and being transparent with my participants helped me to ensure my research included the richest possible, yet also ethically collected, data.

Still, repositioning oneself in the field is not an easy task, as in all contexts, decisions are made purely intuitively and quickly (Fook 1999; Noh 2019). Even when reflecting on them, I sometimes find it difficult to pinpoint the underlying principles that guided my navigation within those relationships, except for the larger ethical considerations and genuine care for my participants. Because of this, I would approach fieldwork again with a checklist to ask myself: "Have I come here as a researcher or a friend?" "Am I allowed to switch my roles?" "Who gave me permission to do so?" and, most importantly, "If I had to face the person who knew that I have used some of the information provided to me under ambiguous circumstances, how would our encounter go?" If the answer was uncomfortable, embarrassing, or similar, I chose not to use the information. Indeed, relationships between a researcher and their informants often become multi-layered and do not neatly fit into either the personal or the professional domain, as well as changing over the course of the research (Pink 2000). Therefore, such a checklist may not be useful for all scholars, and others may also need to ask themselves additional questions to ensure their research is ethical and transparent. Nevertheless, in my experience, adding new dimensions to relationships in the field in no way obscured, blurred, or made more fluid the professional aspect of it. I found compartmentalization very useful for my own relationship management in the field. In particular, it helped me to be more aware of my role and of the actions that the role permits, even if the role(s) were quickly alternating.

Involvement and Detachment from the Fieldwork (Not Quite) at Home

Every experience in the field is different, as is everyone's mental capacity to cope with its challenges. When reflecting on my own field, an important consideration that I believe needs to be further stressed is about whether staying at a site in the field is productive or counterproductive for the progress of the project and for the researcher's well-being. From my experience, making an informed decision about whether to persevere or take a break from research can be critical for the outcome and findings of the research. If successfully navigated, moving between the field and home can even be an asset for some projects as it allows for connections with research participants, but still gives the researcher space and time away from the field. Indeed, I realized that compartmentalizing my roles and the spheres of home and field was also a useful strategy to resolve the challenge of my multiple belongings.

Importantly, taking breaks from the field can hinder the formation of strong and steady bonds with the participants, impede immersion in the field, and subsequently result in poor data collection (Evans-Pritchard 2004[1951], 79). However, this line of thinking assumes that data collection and analysis must be two separate stages of the research that neatly follow one another. In my experience, data collection and data analysis were always, to some extent, iterative and overlapping. For example, after some of my first interviews and observations, I went back to the UK to transcribe and analyze the material. It was only after doing the analytical work that I noticed the topic of trust and safety often surfacing in my conversations. This helped me to attune my focus on these topics during my subsequent visits. As I continued to alternate between data collection and analysis throughout my research, I believe this approach only deepened my understanding of the people I studied. In addition, the data I collected was not limited to the fieldwork as a source, since my participants often contacted me through social media and/or emails when I was away from the field, sending me reflections about our conversations or adding new information they had forgotten to mention while I was recording their stories. Some simply shared their daily lives with me, which helped me to better interpret the stories I had recorded. Thus, in my experience, involvement in and detachment from the field (Powdermaker 1967) came hand in hand during my research process. In addition, I must emphasize that my understanding of my research, field, and even myself came through my experiences being involved in and detached from the field (Till 2011).

Apart from positively reinforcing data collection and analysis, taking breaks from fieldwork also helped me to mitigate the effects of fieldwork-related stress and anxiety. Even under more favorable conditions, I found fieldwork to be a very intense experience, as I was constantly observing and being observed, staying focused, managing my own appearance, taking mental and written notes, and speaking with people to understand the meaning behind their words and actions. This often led to exhaustion and anxiety that I had not expected before my data collection. Indeed, the literature on fieldwork experiences point out that such feelings are common, and the researchers often take breaks to ease them (for more details see Powdermaker 1967; Shore 1999 and Nordquest 2007). Prior to entering the field, I wish I had paid more attention to the literature that pointed out the necessity to "escape the field" (Powdermaker 1967, 100), as it could have provided reassurance and underpinned my coping strategies. It was only later in my program that I discovered some of my colleagues had intuitively found similar approaches to disconnect and mentally rest from their field research, such as by going to the gym and even meeting with a mental health specialist online. As I was doing my fieldwork "at home," I also had many opportunities to retreat from the field while remaining in the field. Because I had family and friends there, I often surrounded myself with people I was able to voice my anxieties to, speaking about my fatigue, or simply watching a movie in a relaxed and familiar environment. Not only did this help me to manage my stress but also to sustain a sharper mind in the field. While immersing oneself in the field is essential for understanding the studied phenomena, these opportunities showed me the importance of taking a break from research when overwhelmed by feelings of loneliness, anxiety, anger, or the "need for an escape" (Powdermaker 1967, 100). Had I been forced to stay in the field, I believe it may have impeded my ability to collect and adequately evaluate my data, as well as risked damaging relationships with participants in the field.

Evaluating the Risks of Fieldwork

I was encouraged to think about the ways to ensure the psychological and physical well-being of both my research participants and myself. In order to receive approval from the University Research Ethics Committee, I had to outline the possible risks that might affect the safety of my research participants as well as my own safety. However, at that time I thought that the major risk of the research was related to data confidentiality (i.e., not exposing research participants' identities) and making sure that my participants felt safe and comfortable during our encounters (Cook 2012). As I was concerned about exposing the research participants' identities, I never approached potential research participants directly but used prior contacts and online LGBTQ groups on Facebook to spread the word about my research, and I only recruited the participants who expressed interest in the research. Prior to our meetings I used to send information to the participants about the nature of my research and how I would use the data and encouraged them to share any concerns regarding it. Before starting the interview, I would again explain the purpose of the project and how the data would be used. In addition, I made it clear that the research participants could choose to withdraw from the project at any time before publication of the study.

When it came to the risks related to me, I mostly thought of the ways to protect myself from physical danger. As I did not personally know the majority (apart from three) of my research participants prior to our interview meeting, I thought that there might be a slight risk of being attacked and/or raped by someone pretending to be a research participant. In fact, one of the interviewees expressed the same concern after our interview by saying: "I admire your confidence to come and see me all by yourself. For all you knew I could have been a serial killer." I attempted to reduce the risk by asking the participants before our meeting how they found out about the research. I considered it to be reasonably safe to meet if they mentioned another research participant, or a private online LGBT group as their source. In addition, I was telling the city and the time of an interview to a trusted person and instructed them to contact the police if they did not hear back from me in 10 hours. Fortunately, I did not experience any dangerous situations in the field.

As I reflect on my PhD fieldwork, I think that I did not give sufficient attention to the psychological impact that the fieldwork would have on me. I primarily thought about reflexivity (such as keeping reflexive diaries and audit trails throughout the research process) as a precautionary measure, which ensures the transparency of the research process. I did not think of it as a necessary measure to prepare myself for handling the day-to-day experiences in the field. But as my experiences demonstrate, a researcher's agenda in data collection and analysis is an inherent part of fieldwork (Charmaz and Belgrave 2012). Despite the recent developments around researchers' positionalities in the field and the importance of reflexivity, I was driven by the orthodox view to keep separate the field and the researcher's personal circumstances. One of the important lessons that I learned during the fieldwork was the importance of fieldwork diaries not only to ensure transparency and rigor of the research process but also to help the researcher to make sense of their experiences in the field.

Conclusion

As this chapter has shown, I found myself in several positions during the course of my fieldwork. In terms of my nationality, I perceived myself as a homecomer (Schuetz 1945), as someone who partially finds the environment familiar, while certain aspects of the environment still set me apart from the local people. My research participants, on the other hand, saw me as an insider and a fellow Lithuanian. In terms of social class, I felt like an "interloper" (Ryan and Sackrey cited in Johansson and Jones 2019, 1528), marginally belonging to both the working and middle classes. However, in the eyes of my informants, I was primarily perceived as a struggling migrant and only secondarily as an aspiring academic. The shared knowledge of the particularities of the migrant experience, also fostered my insider position and helped me to establish a good rapport and gain access in the field, which was further strengthened by a shared sexual identity. Nevertheless, there were times during my fieldwork when I repositioned myself as an outsider in terms of my sexuality, as identifying as part of the researched group. This made it exceedingly difficult for me to engage objectively with the data. My fieldwork experiences therefore reinforce the complex dynamics requiring navigation when researchers' and participants' identities overlap, as well as the necessity to be reflexive, including even dissociating from research participants when need be.

As I conducted fieldwork in my "home" country, I was therefore switching between the roles of a friend and a family member while staying outside of the field, and that of a researcher while in the field. By doing so, I tried to protect myself from difficult circumstances related to the field (Till 2011). In this way, alternating between personal and scientific-objective modes of engagement (Nordquest 2007) helped me to sustain focus and constantly reevaluate the situations occurring in the field. As my research progressed, my role in the field also shifted from being a researcher to a friend to my participants, which fundamentally brought up other questions of informed consent in data collection and the need to protect newly established relationships.

During the fieldwork, I therefore experienced multiple belongings and positions and found that repositioning and reordering *stable* positions were key to successful progress. While I agree that research experiences are often much more complicated than the assumed simple separation between the home and the field, I would still define the relationship between my home and field (and between my personal self and my role as a researcher) as interspersed and *inter*changing, rather than constantly changing. In fact, during my fieldwork, I cognitively and geographically compartmentalized my researcher and personal selves, as well as my home and my field (Nordquest 2007) to maintain an ethical engagement in the field and to look after my well-being. I found it necessary to do this as I was aware that my different roles and relationships (personal and professional alike) come with different obligations and expectations. Thus, I sought to ensure these expectations were met as a failure to do so could have damaging effects in both spheres, for my participants as well as for myself.

Notes

1 ILGA-Europe is the European Region of the International Lesbian, Gay, Bisexual, Trans and Intersex Association. The information about the LGBTI situation in Lithuania can be accessed at https://rainbow-europe.org/#8644/0/0.

2 Grounded theory consists of "systematic, yet flexible guidelines for collecting and analysing qualitative data" (Charmaz 2014, 1) to construct a theory from the collected data. Hence, the theory is "grounded" in the data. I employ a constructivist variant of grounded theory (Charmaz 2014), which sees theory building as informed not purely by the data but also by researcher's "social, epistemological, and research location" (Charmaz and Belgrave 2012, 348).

3 According to official statistics, in the period of 2010–2020 alone, nearly 500,000 Lithuanians migrated abroad, which is a significant number for a population of 3 million in 2010 (Official Statistics Portal Lithuania 2010).

References

Amit, V. 2000. "Introduction." In *Constructing the Field. Ethnographic Fieldwork in the Contemporary World*, edited by V. Amit, 1–18. London and New York: Routledge.

Bukodi, E. and J. Goldthorpe. 2018. *Social Mobility and Education in Britain: Research, Politics and Policy*. Cambridge: Cambridge University Press. DOI: 10.1017/9781108567404.

Cederberg, M. 2017. "Social Class and International Migration: Female Migrants' Narratives of Social Mobility and Social Status." *Migration Studies* 5 (2): 149–167. DOI: 10.1093/migration/mnw026.

Charmaz, K. 2014. *Constructing Grounded Theory*. 2nd edition. Thousand Oaks: SAGE Publications.

Charmaz, K. and L. Belgrave. 2012. "Qualitative Interviewing and Grounded Theory Analysis." In *The SAGE Handbook of Interview Research: The Complexity of the Craft*, edited by J. Gubrium, J. Holstein, A. Marvasti and K. McKinney, 347–366. Thousand Oaks: SAGE Publications. DOI: 10.4135/9781452218403.n25.

Cook, K. 2012. "Stigma and the Interview Encounter." In *The SAGE Handbook of Interview Research: The Complexity of the Craft*, edited by J. Gubrium, J. Holstein, A. Marvasti and K. McKinney, 333–345. Thousand Oaks: SAGE Publications.

Ellis, C. 1995. "Emotional and Ethical Quagmires in Returning to the Field." *Journal of Contemporary Ethnography* 24 (1): 68–98. DOI: 10.1177/089124195024001003.

Evans-Pritchard, E. 2004[1951]. *Social Anthropology*. London: Routledge. DOI: 10.4324/9781315017655.

Fook, J. 1999. "Reflexivity as Method." *Annual Review of Health Social Science* 9 (1): 11–20. DOI: 10.5172/hesr.1999.9.1.11.

Gardner, K. 1999. "Location and Relocation: Home, 'The Field' and Anthropological Ethics (Sylhet, Bangladesh)." In *Being There: Fieldwork in Anthropology*, edited by C. Watson, 49–73. London: Pluto Press.

Given, L. 2008. *The Sage Encyclopedia of Qualitative Research Methods*. Thousand Oaks: SAGE Publications. DOI: 10.4135/9781412963909.

Hellawell, D. 2006. "Inside-Out: Analysis of the Insider-Outsider Concept as a Heuristic Device to Develop Reflexivity in Students Doing Qualitative Research." *Teaching in Higher Education* 11 (4): 483–494. DOI: 10.1080/13562510600874292.

ILGA-Europe. 2022. *Annual Review of the Human Rights Situation of Lesbian, Gay, Bisexual, Trans and Intersex People in Europe and Central Asia*. Available at: https://rainbow-europe.org/annual-review (accessed on March 5, 2022).

Johansson, M. and S. Jones. 2019. "Interlopers in Class: A Duoethnography of Working-Class Women Academics." *Gender, Work, and Organization* 26 (11): 1527–1545. DOI: 10.1111/gwao.12398.

Katyal, K. and M. King. 2011. "'Outsiderness' and 'Insiderness' in a Confucian Society: Complexity of Contexts." *Comparative Education* 47 (3): 327–341. DOI: 10.1080/03050068.2011.586765.

Knowles, C. 1999. "Here and There: Doing Transnational Fieldwork." In *Constructing the Field. Ethnographic Fieldwork in the Contemporary World*, edited by V. Amit, 54–70. London and New York: Routledge. DOI: 10.4324/9780203450789.

McNess, E., A. Lore and M. Crossley. 2013. "'Ethnographic Dazzle' and the Construction of the 'Other': Revisiting Dimensions of Insider and Outsider Research for International and Comparative Education." *Compare: A Journal of Comparative and International Education* 45 (2): 295–316. DOI: 10.1080/03057925.2013.854616.

Milligan, L. 2016. "Insider-Outsider-Inbetweener? Researcher Positioning, Participative Methods and Cross-Cultural Educational Research." *Compare: A Journal of Comparative and International Education* 46 (2): 235–250. DOI: 10.1080/03057925.2014.928510.

Munthali, A. 2001. "Doing Fieldwork at Home: Some Personal Experiences among the Tumbuka of Northern Malawi." *The African Anthropologist* 8 (2): 114–136. DOI: 10.4314/aa.v8i2.23107.

Noh, J. 2019. "Negotiating Positions through Reflexivity in International Fieldwork." *International Social Work* 62 (1): 330–336. DOI: 10.1177/0020872817725140.

Nordquest, M. 2007. "Of Hats and Switches: Doing Fieldwork 'at Home.'" *North American Dialogue* 10 (1): 18–20. DOI: 10.1525/nad.2007.10.1.18.

Official Statistics Portal Lithuania. 2010. "International Migration Flows." Available at: https://osp.stat.gov.lt/EN/statistiniu-rodikliu-analize?hash=7ae84706-e252-40de-85d8-35baecbbb038#/ (accessed on July 20, 2021).

Owton, H. and J. Allen-Collinson. 2014. "Close But Not Too Close: Friendship as Method(ology) in Ethnographic Research Encounters." *Journal of Contemporary Ethnography* 43 (3): 285–305. DOI: 10.1177/0891241613495410.

Pink, S. 2000. "'Informants' Who Came Home." In *Constructing the Field. Ethnographic Fieldwork in the Contemporary World*, edited by V. Amit, 96–119. London and New York: Routledge.

Powdermaker, H. 1967. *Stranger and Friend. The Way of an Anthropologist*. London: W. W. Norton and Company.

Saar, M. and E. Saar. 2019. "Can the Concept of Lifestyle Migration Be Applied to Return Migration? The Case of Estonians in the UK." *International Migration* 58 (2): 52–66. DOI: 10.1111/imig.12601.

Schuetz, A. 1945. "The Homecomer." *American Journal of Sociology* 50 (5): 369–376. DOI: 10.1086/219654.

Sen, A. 2020. "Three Evils of Citizenship Education in Turkey: Ethno-Religious Nationalism, Statism, and Neoliberalism." *Critical Studies in Education*. DOI: 10.1080/17508487.2020.1761849.

Shore, C. 1999. "Fictions of Fieldwork: Depicting the 'Self' in Ethnographic Writing (Italy)." In *Being There: Fieldwork in Anthropology*, edited by C. Watson, 25–48. London: Pluto Press. DOI: 10.2307/j.ctt18fs9h3.5.

Sultana, F. 2007. "Reflexivity, Positionality, and Participatory Ethics: Negotiating Fieldwork Dilemmas in International Research." *ACME: An International E-Journal of Critical Geographies* 6 (3): 374–385.

Till, K. 2011. "Returning Home and to the Field." *Geographical Review* 91 (1–2): 46–56. DOI: 10.1111/j.1931-0846.2001.tb00457.x.

Tillmann-Healy, L. 2003. "Friendship as Method." *Qualitative Inquiry* 9 (5): 729–749. DOI: 10.1177/1077800403254894.

Zhao, Y. 2017. "Doing Fieldwork the Chinese Way: A Returning Researcher's Insider/Outsider Status in Her Home Town." *Area* 49 (2): 185–191. DOI: 10.1111/area.12314.

3 A Woman of Her Word Prepared for the Worst

Researching Drug Trafficking in Kazakhstan[1]

Zhaniya Turlubekova
Independent researcher

Introduction

On October 7, 2020, the president of Kazakhstan, Kassym-Zhomart Tokayev, signed a set of amendments strengthening anti-corruption measures in Kazakhstan. The amendments criminalized the provocative actions of operatives and other state officials engaged in anti-corruption criminal cases. The same legislation also criminalizes the acceptance of material gifts received by state officials and members of their families, as well as prohibiting the employment of relatives in the same governmental agencies (Kazinform 2020). This change occurred only two months after my doctoral thesis defense and very much lifted my spirits as it addressed many of the central issues that I had studied and written about in my doctoral dissertation. Nevertheless, I had a bittersweet feeling of regret when these measures came into effect, as my field research would have been much more comfortable had these changes preceded my entry into the field.

While fieldwork can be perceived as one of the most important professional experiences in academia, especially in the social science fields of anthropology and sociology, as a young female academic, I wish I had been better prepared for the unexpectedly complex processes that I faced while in the field. Technically, I was conducting research at "home" and like other Central Asian female scholars before me (Du Boulay 2019), I was certain that I was familiar with, if not already embedded in, the local context. However, what constitutes the "field" and our "homes" is a problematic distinction, as returning to the country one was born in to do fieldwork does not always actually mean returning "home." In my situation, when I returned to my home country to conduct research, I was not only entering a country I had a prior connection to but also a previously unknown territory: the world of drug trafficking.

In my doctoral research, I explored how Kazakhstan, as the closest and perhaps most important neighbor to Russia, has been affected by the flow of drug trafficking to Russian drug markets. I analyzed how national legislation, anti-corruption measures, and the state's geographic proximity to Russia affected marijuana and heroin trafficking in Kazakhstan

DOI: 10.4324/9781003144168-5

(Turlubekova 2020). Importantly, I was familiar with Kazakhstan's criminal code and criminal procedural legislation from my previous studies. Moreover, I knew that state drug policies are an important factor in international relations, particularly as former colonial masters and regional hegemonies have been accused of influencing national drug policies of their allies, and wider discussions about global drug policy (Bullington and Block 1990; Weitz 2006; Utyasheva et al. 2009; Mercille 2011; Paoli, Greenfield and Reuter 2012; Patten 2016; Tinasti and Barbosa 2017; Lopega 2019). For example, Utyasheva et al. (2009, 78) described Eurasian and Russian drug policies as highly prohibitive, coercive, and "raising serious human rights concerns." The authors also argued that the Kazakhstani parliament adopted one of the toughest sanctions for drug-related offences in Eurasia which introduced life imprisonment for drug trafficking in cases of extra-large quantities of drugs, trafficking by organized groups, and selling drugs to minors (Utyasheva et al. 2009, 100).

Given the challenging topic that I study, prior to entering my field, I identified several reasons why my research was considered "high-risk." Most obviously, the research was expected to take place within the context of corruption, organized crime, possible terrorist funding and activities, and politically sensitive environments (for more, see Abdirov 1999; Cornell 2005; Golunov and McDermott 2005; Golunov 2007; Mohapatra 2007). Since the drug dealers and drug traffickers that I interviewed maintain low profiles due to the high-risk and illegal nature of their activities, they were, unsurprisingly, very challenging for me to access (De Danieli 2014). Due to my respondents' criminal activities, I also faced numerous methodological challenges and ethical dilemmas while conducting my research, as well as physical and professional risks which are well known to scholars in criminology (see, for example, Ferrell and Hamm 1998). Many of these are explored throughout this chapter.

But while ethnographic research on drug trafficking is indeed arduous and costly, it can produce remarkable results. Such examples include publications by Adler and Adler (1983), who spent almost six years collecting ethnographic data about high-level drug traffickers operating in a city close to the Mexican border; Zaitch (2002), who demonstrated that the cocaine business is mostly populated by small, flexible, and to a large extent ad hoc units, in sharp contrast to the popular image of "cartel-like" operations; and Lalander (2003), who looked into the world of young heroin users. Although drugs are often associated with drug violence, Omelicheva and Markowitz (2019) observed in their research that the Central Asian drug markets are characterized by a low level of drug violence. My research setting, therefore, did not correlate with the popular stereotype portraying drug cartels' wide use of excessive violence in their operations.

But while my research differs from that described in the previous literature, I was aware that I might encounter some unique challenges related to my context of the former Soviet region. To prepare for fieldwork, I therefore acquainted myself with the criminological research on organized crime and the Russian mafia in other post-Soviet countries (Varese 2001, 2017; Slade 2013; Volkov 2016). Furthermore, I knew that as a young Kazakh woman researching drug trafficking in Kazakhstan, where the nation-building discourse firmly cemented men's position of dominance over women (Kudaibergenova 2018; Arystanbek 2020), I would attract a lot of unwanted attention and possibly face additional risks to my safety and security. Yet, I was inspired by previous groundbreaking research by other female researchers on the Russian police system (McCarthy 2016), Russian prisons (Piacentini 2013), and the Russian drug market (Paoli 2001). These female scholars have demonstrated that women can safely and ethically conduct research on crime-related issues in risky, male-dominated situations within post-Soviet societies. However, as a young early-career scholar, I did not have the same experience as these more established scholars. For this reason, and despite preparing myself with the help of the abovementioned literature and other studies related to fieldwork methodologies, my engagement with the field was more challenging than I had anticipated. Although this was partly because I am an early-career scholar and female Kazakh, it was more closely connected to my sincere motivation to help and protect my respondents to the best of my abilities during my fieldwork. In the following pages, I therefore critically reflect on how different aspects of my identity, specifically my socioeconomic class, age, and gender, as well as the political changes in Kazakhstan, influenced the progression of my fieldwork.

Unexpected Political Turmoil in the Field

Fujii (2015) explains that the importance of our field observations lies not only in the stories they tell us but also in what they suggest about the larger political and social world within which they (and the researcher) are embedded. In this way, our research is influenced and constrained by the social and political systems of our field sites, including the developments which occur within these environments (Sultana 2007, 381). In my experience, my research coincided with, and was thus constrained by, several massive changes outside of my control. The overwhelming nature and unexpected duration of the campaign against corruption, the changes in legislation, and the structure of the criminal justice system indicated a possible mismatch between the Kazakhstan I knew and the Kazakhstan I had to do research on.

In particular, when I entered the field in 2017, the Kazakhstani government was implementing new measures to address police corruption on a massive scale. While I spent 43 weeks conducting fieldwork between January 2017 and April 2018, not a single week passed without at least one headline on the scandalous arrest of a corrupt police officer or state official (Turlubekova 2020). Naturally, the intensity, suddenness, and extent of the state's anti-corruption campaign was met with hostility, skepticism, and anger within internal police hierarchies. Luc Duhamel's (2010) work on Andropov's KGB (*Komitet gosudarstvennoj bezopasnosti)* anti-corruption campaign helped me to develop a broader picture of the fear, disbelief, and resentment of police officers toward the state-led anti-corruption measures that I observed during my fieldwork in Kazakhstan. Specifically, Duhamel describes the anti-corruption measures taken by the KGB in the late 1980s, showing how corruption can endanger the existence of a regime. He also traces how KGB officials were recruited and raised as radically committed patriots who traveled great distances and did not shy away from exposing corrupt state officials (Duhamel 2010). Indeed, when I began my fieldwork, the local security services were very preoccupied with the fight against corruption. Some of my future respondents even approached me and my work with hostility and suspicion, perhaps because they were mistrustful of coincidences: it seemed to some that my arrival in the field was connected with the recent initiation of anti-corruption measures. At the time, I was unaware of this coincidental timing because none of the academic sources I had engaged with had prepared me to recognize the early signs of the crackdown by the security services aimed at eradicating corruption. Hence, I instead mistakenly thought people were confused by a confident but shy, petite, and young Kazakh scholar who was affiliated with a Western university. In any case, these early interactions very much influenced the outcome of my PhD fieldwork.

As my research topic was related to themes considered to be "high-risk," I prepared extensively for my field research. Prior to entering the field, I read widely about the role and practice of ethnographic research in criminology. My supervisory team also evaluated and approved my field research strategy. I was also encouraged to follow the British Society of Criminology's (BSC) Statement of Ethics (2015) and obliged to comply with European Code of Conduct for Research Integrity, the Netherlands Code of Conduct for Research Integrity, and Utrecht University Research Integrity regulations. Like Decker, Decker, and Van Winkle (1996), I knew that the local police had the legal right to demand disclosure of any information about planned or committed criminal activities. Thus, I included a paragraph in my information sheet to participants that explicitly stated this legal requirement and a sentence explaining I was not seeking information that could be categorized as such, or any information that could endanger them, or anyone else. On occasions when people disregarded my informed consent form, I would orally explain it to them and highlight the fact that doing no harm was at the top of my list of research priorities. To ensure my research was ethical and to further gain the trust of my participants, prior to every interview, I mentioned that I did not work for the police, KNB (*Komitet nacional'noj bezopasnosti*) (Kazakhstani heir of the Soviet KGB), or any governmental institutions and that I was only interested in information that they were comfortable sharing with me. I also made it very clear that they could decline to answer any question and withdraw their participation at any time. After explaining that avoiding harm to them was one of my main priorities, my participants usually seemed more relaxed and reassured. In fact, some interlocutors almost seemed too relaxed, to the extent that I sometimes felt the need to stop recording the conversation to protect the confidentiality of my participants. In these instances, I asked the participant whether he or she would still want me to record the particular information or delete it from the record.

While protecting my participants was of utmost concern, I observed various reactions to my protective behavior. For instance, some respondents laughed about my informed consent form and my cautious nature, making comments like: "you did become foreign, young lady." Others almost proudly replied that they "had already paid" for their actions and that the local prosecution offices and journalists had already made this information public. Luckily, most of the themes we discussed were related to public information about crimes that had been already processed by the Kazakhstani Criminal Justice system. My priority to put my interlocutors' security and confidentiality before my own academic outputs helped me to gain their trust and to change their perception of me as a possible foreign or national intelligence agent. Moreover, as I am trained as a lawyer, I made sure that the questions I asked under no circumstance reminded my informants about what opposing institutions expected or hoped to learn from them. I therefore formulated the questions as open-endedly as possible to allow them to speak freely – a decision that really helped me to establish a fragile rapport and trust.

Despite my precautions, I soon came to understand Till's (2001) claim that it is impossible to know in advance how our understanding of ourselves may change over the course of our research as we move back and forth between our homes and fields. In fact, there remain many possibilities for how our fieldwork experiences may play out; for example, discrepancies may exist between the image we have of our positionality as a researcher and how our respondents perceive us (Milligan 2016), with our respondents even placing us into predefined, assumed identity categories that differ substantially from our own self-identification (Noh 2019). In my experience, and because of my ascribed identities as a "Western criminologist," "theoretical scholar," and female civilian, I was never certain if my participants saw me as an "insider" or an "outsider" while in the field. In the following sections, I therefore aim to shed light on these three aspects of my identity and the ways they influenced the progress of my data collection in the field.

Do *No Harm*, Mean *No Harm*

As Milligan (2016) suggested, the active choices of researchers in their research designs and data collection can very much affect how they are perceived by their research participants. Yet, it is not solely the researchers' recruitment and data-collection strategies which shape their interactions with participants but also their underlying belief systems. Although I had not identified myself as a "critical criminologist" (Sykes 1974, 208) prior to my fieldwork, I shared Sykes' (1974) understanding that both criminal legislation and its enforcement are designed by one class to control another. When entering the field, I thus genuinely hoped to bring some justice – or at least some attention – to the struggles of my respondents, as I naively believed that these people needed help to liberate themselves from the mistreatment by their superiors and the severity of the criminal justice system within which they lived. Similar to Becker (1967) and Mamadshoeva (2019), I hoped that I might better understand my interlocutors, learn the way they saw the world, and allow them to express their truths. To ensure my research was ethical, I therefore decided to follow Sultana's (2007) principle of causing "no harm" to my respondents and, instead, tried to make their often-silenced voices heard.

Yet, my experiences demonstrated that what occurs at grassroots level is often much more complicated than we originally realize. My attempt "to give voice to the oppressed" (Mamadshoeva 2019), as well as my wish to actively "avoid any harm" to my participants, had a profound effect on the way I was perceived by my respondents in the field. In particular, it "othered" me in the eyes of most respondents. Like Mamadshoeva (2019), who was appalled by her respondents' seeming acceptance of gender-based violence in Tajikistan, I was sometimes surprised or even shocked by my respondents' perspectives on the effects of drug policy and criminal law in Kazakhstan; many of my respondents had already adjusted to the regime's normalized (occasionally admired) practices and policing solutions. Thus, similar to Mamadshoeva (2019), I often found myself wondering if (and how) some of my respondents saw abuses of power (in Mamadshoeva's case, this was the general public's toleration of sexual abuse) and accepted it as the only feasible way to live. Since my idealistic approach to the field, law, and criminal justice led to comparisons with Pier Bezukhov of Tolstoy's *War and Peace*, I accordingly had difficulties making sense of my own relation to and positionality vis-à-vis my respondents because many of them seemed to accept what I saw as the devastating consequences of a drug policy. Yet, my participants often noted how naive I was, laughing about my questions and my hopes of bringing more justice or even improving the criminal justice system with my research. They drew attention to my innocence by calling me a "little girl."

Moreover, when I asked about drug trafficking in Kazakhstan, my respondents instead regularly described Russian drug policies and explained to me that I must be "one of those too liberal, democratic drug policy supporters" from the West. This was sometimes after less than five minutes of the conversation, during which I barely had a chance to introduce myself. Surprisingly, in such interviews, I noticed admiration toward American drug policies in terms of their efficiency but, in general, I was not comfortable being placed in the category of "these democratic drug policies supporters" because, as a criminologist, I was well informed about the devastating effects of America's so-called "War on Drugs." As I believed I had nothing to do with my participants' fear of ideological diversion that could result from complying with Western drug policies, I therefore often felt annoyed, or even angry, when my participants labeled me as "yet another Western/theoretical scholar." In fact, I can recall several instances when potential respondents suggested that I write "whatever [my] boss wants to hear," rather than agreeing to an interview with me. Although I became accustomed to hearing similar sentiments, the first time left me with feelings of confusion, and even devastation, because, as a young qualitative criminologist, I thought that it was my socialization or rapport-building skills that had made them behave this way.

But as I realized over the course of my fieldwork, my respondents' aversion to speaking with me seemed to stem from their local mentality of familiarity with adjustive information discourses, rather than my personality as a "liberal, democratic drug policy supporter." Their suspicion toward my research was also an understandable consequence of surviving in a highly idiosyncratic information discourse. In fact, "information blockade was an indispensable part of Soviet ideological indoctrination" (Khazanov 2008, 134) and its legacies are still noticeable. In essence, these people were raised in a world of controlled media, corruption scandals, statistical evaluation of performance ordered by party officials, state budget embezzlement, and harsh Soviet laws. Moreover, the tradition of barely verifiable information discourses has been entrenched much deeper – what are recognized today as Soviet propaganda campaigns were once taught and presented as policy achievements in their history classes (Andrieu 2011). My participants' jibes about writing "whatever your boss wants to hear" could therefore be interpreted as a positive sign, as I understand in retrospect; instead of politely sending me away, they subtly indicated that I had finally entered the field and was turning into a trusted outsider.

I soon began to understand that in order to bring about more justice for my respondents, and to facilitate deeper and richer reflections from them, I had to take them seriously "on their own terms" (Mamadshoeva 2019) and to avoid any "ulterior motives other than a deep desire to understand (...) how they saw the world" (Varese 2017, 7). Overall, as an inquisitive introvert, I was very happy to quietly listen to my high-achieving respondents. In essence, their professional activities were extraordinary: drug traffickers and corrupt police were taking risks by messing with national security interests. Female drug entrepreneurs managed their businesses admirably, while honest police officials and less influential retail drug dealers faced ethical challenges ranging from assessing the trustworthiness of colleagues and clients to searching for ways to function effectively – ideally within legally permissible limits. Collectively, these experiences fascinated me. As I was an observer in their world and remained silent even when I personally disagreed with their means, views, and solutions, I was still always impressed by their bravery, decisiveness, and survival in their illicit business environments. Although my genuine belief in the presumption of innocence and the rule of law was often mocked by my informants, it was because of my inquisitive approach and desire to learn about their world that a great number of my respondents were rather amused to speak with me.

The protocol designed to ensure my personal safety was relatively simple. Most importantly, as I had returned to my home country to conduct fieldwork, I therefore, by definition, was embedded in circles and communities of people I knew. Therefore, I would tell a trusted person or a family member where and what time I planned to meet one of my interviewees. I was often driven or picked up from interview locations by friends, family members, and other acquaintances. Most of the time, I scheduled a meeting with a friend in a café directly after my interview. I went there directly, made my notes, drew a mind map supporting the interview notes, and waited for the arrival of a friend. In addition, I did my best to call them on my way to and from the interviews so that they knew exactly where I was. This protective measure was suggested by other scholars working in high-risk environments (see, for example, Decker et al. 1996). It has also become something that many women traveling alone regularly do to ensure their safety. I also carried two mobile phones with me in case of the very unfortunate scenario that my smartphone would not survive the Kazakhstani winter or be stolen. These measures made me feel more comfortable at my field sites while also ensuring my safety.

Shared "Femininity"

Importantly, apart from feeling pushed into two categories – the Western "otherness" and as a trained "theoretical scholar" studying at a European university, I noticed that my respondents, and especially less privileged drug entrepreneurs, also "othered" me in various ways. Specifically, and similarly to Sultana (2007), my respondents assigned a certain geopolitical identity to me because I was tied to a Western higher educational institution. Another local female scholar, Du Boulay (2019), also highlighted the weight of educational, class-related differences when conducting research, which very much resembled my own experiences. Like Du Boulay (2019), I studied at an elite school, enjoyed tutoring foreign languages, history, and legal theory and had attended a specialized drawing school since my early childhood. Like with Du Boulay (2019), it was through my research, specifically becoming aware of the disproportional impact of the drug trafficking on less-privileged young Kazakhstani women, that I really came to notice my own privileged background. I suddenly understood the relative insignificance of my "luxury problems" of being objectivized, isolated from the outside world behind the medals and honors awarded to male relatives for exemplary legal work or military service, or even my data-collection challenges.

One of these challenges was my recruitment of female drug dealers. As I recruited participants for my research, I tried to initiate conversations with female sex workers as I had read that some of them might either be regular customers of drug dealers, exchange services with them, or start dealing drugs themselves (see Hunt 1990; Young, Boyd and Hubbell 2000; Romero-Daza, Weeks and Singer 2003). Although I thought they might be able to offer valuable information for my research, they often declined my requests either with hostility or silence. On one occasion, a female worker did begin chatting with me, but then sharply shifted the conversation to the cost of my shoes, before asking where I styled my hair. I joyfully shared the details of my hairstylist, but then she suddenly turned and left. A few days later, another informant commented that I was sabotaging my own safety by engaging in such conversations "because local junkies might find out how much your earrings cost." Hence, I quickly came to realize that my "shared femininity" (Malyutina 2014, 122) with female sex workers did not help me to establish relationships with them because my clothing and hairstyle clearly identified me as an outsider, as someone belonging to a different social stratum. While I thought I was dressed appropriately to conduct my research, as I learned through these experiences, I had to reassess my approach to managing impressions when engaging with particular groups of people.

As time passed, I therefore began to understand Thibault's (2021) argument that despite our best attempts, female scholars will often be judged on their looks and sexual availability. Although I dressed conservatively while in the field, I learned that my macho-like respondents still sometimes described me as "a pretty piece of academia," "a baby-scholar with a nice voice and good legs," and "all-proper, but juicy and well-mannered." I often felt embarrassed by such descriptions because I wanted them to give their attention to my study, not to me, and preferably through the use of other words. Still, I was prepared to deal with sexist comments toward me and other women scholars in the masculine environments that I study, and, like Adams (1999, 346), I also observed that being introduced as a "delightful/beautiful girl" actually worked well in predominantly male-dominated territory. Indeed, I still certainly felt like I belonged to a gender minority at times while conducting my fieldwork (Turlubekova 2020).

But whereas being a young Kazakh woman was less helpful during interviews with female sex workers, I also discovered that my gender actually helped me to connect with notorious female drug entrepreneurs. My "shared femininity" (Malyutina 2014, 122) with these women enabled us to form a bond and allowed me to explore the drug business and emotional side of their largely male-dominated professional lives. Because of this, I listened and engaged with them regularly, often hearing their stories about betrayal by peers, partners, and superiors, which showed me the extent of the connections between those involved, and the reliability of the protection (or lack thereof) offered by corrupt police. Throughout these interactions,

I also came to see how their fears and achievements revealed risk management strategies; in particular, their regrets and aspirations shed light on the effects of drug policy and anti-corruption measures (Turlubekova 2020).

You Are a So-Called Female Drug Lord, Aren't You?

Similar to ethnic minorities in Western Europe (Paoli and Reuter 2008), the enormously brave and decisive women I met while in the field had entered the drug business mostly because they had been routinely denied access to the legitimate economic activities in other sectors. I came to this conclusion after I heard numerous stories about younger women imprisoned on drug-trafficking charges as a result of exploitation by their husbands, boyfriends, or male business partners. In line with Campbell (2008), this finding underlined that drug trafficking can heavily affect women's lives, often more than men's, and regularly leads to female victimization. More importantly, however, is that drug trafficking can, paradoxically, act as a vehicle for female empowerment. The growing feminization of drug trafficking, on the one hand, exposes women to drug violence and manipulation into cooperation by lovers, spouses, and other relatives, but it can also offer liberation from other forms of male control, and a source of adventure and excitement (Campbell 2008). As was revealed to me throughout my studies in Kazakhstan, some women entered the market because they suddenly found themselves as the main providers in their families or because their loved ones could no longer manage their own drug addictions. Others had entered the business because their own drug consumption progressed and demanded further engagement. Yet, despite their differences, my participants all had something in common: they all suffered as a consequence of drug policies crafted by decision-makers who seemed to be blissfully unaware of how the drug policy and its enforcement were affecting the people at the grassroots.

Despite all the challenges, it is noteworthy that I observed several women become relatively successful in this enormously demanding business area. In fact, I met one woman who had been the main manager of a large heroin distribution enterprise for more than a decade – a considerable achievement weighing up the challenges of the early transition period when she founded her business. After the bankruptcy of her small construction company, separation from her husband, and serious illness of her mother, she began to look for a new profitable business opportunity and eventually ended up in the drug-trafficking sector. As she described her professional regrets and achievements, I came to admire her bravery, decisiveness, and strength, which had enabled her to survive in such a demanding business for a relatively long time. More importantly, she had gained considerable experience in managing a business in a very uncertain and unstable environment both with, and against, police. Her devotion to her children, her survival skills in an illegal business, and her relationships with her employees harmoniously mixed into a portrait of an immensely strong and decisive woman.

My interview with her was accordingly anything but the exploitation typically suggested by the scholar-research subject dichotomy. First, she immediately shifted the power-dynamics by using *Ty* instead of a more formal *Vy*.[2] At this point in the interview, we had not developed a familiar relationship, but it is likely that my age played a role in her response to me because, at the time of the interview, she was approximately 25 years my senior and I was fairly convinced that she took me for a fool. Fortunately, I was trained well enough to remain silent during the course of the interview, even though I suspected that she was trying to use me as a valuable addition to her defense strategy until she realized that I was not a journalist. In fact, there was a short while when I felt that she was actively ignoring my questions and pushing her own agenda. In addition, it did not take a lot of effort for me to notice signs of neutralization techniques in her answers. This is demonstrated by the following excerpt:

> Informant: "[we were] going to *banya* [Russian for sauna or Baths] together... That was the type of organized criminal group we had! There was nothing like this. What can I say? That I had runners? I had runners! She had runners too. And we were connected before we knew it. We could not even understand how it was possible in our times. At **that** time [emphasized with a change of tone] it was possible. You get in, but you will never get out."
>
> Author: "At that time?...What time do you mean?"
>
> Informant: "The times of *Cheka*![3] Those days nobody could open their mouth. We thought that we will explain everything, prove our truth... However, we could not..."

Despite being familiar with her history, I did not know what she meant by proving her "truth." I had followed her career closely, watched all press releases where she was mentioned, read every newspaper article about her, and analyzed the testimonies in hundreds of pages of trial transcripts – some of which included very detailed judicial deliberations related to her intercepted phone calls – and yet, this statement made me realize that I did not know her at all. In fact, I quickly realized that I only knew representations and depictions of her. Although I was neither conducting a police inquiry nor collecting evidence against her (or anyone else she was cooperating with) – I came to understand that my task was instead to put everything I *thought* that I knew about her drug business and corruption aside and listen to what she was telling me – she actually challenged almost everything that I *thought* I knew about her. Moreover, as the interview progressed, she became more open and I began to appreciate the precision of her insight into legal analysis. Because of my legal training, this was thus possibly the most challenging and important interview that I had, although it was not until the data analysis stage that I began to fully appreciate the importance of her comments; I continue to be amazed with how informative as, both when she answered enthusiastically and when she remained . More importantly, as the years pass and my knowledge of drug-d problems grows, so does my appreciation of her willingness to sup-the efforts of a young and hopeful scholar.

Honey Trap?

While my interactions with some female participants, like the aforementioned participant, were eye-opening, those with my male participants were equally interesting. My male respondents and I often joked about the National Security Committee, other foreign intelligence establishments, and their imagined interest in my research. But like Sultana (2007), I often felt "othered," or studied, if not provoked, by my respondents. This was abundantly clear in some interactions, such as when one male respondent told me, disapprovingly, that few women of my age have an interest in my research areas – he seemed to assume that crime-related issues, especially serious forms of crime like drug trafficking and organized crime, are exclusively male territories and should not be studied by females.

Still, like Johnson (2009), I noticed that gender differences sometimes helped me to engage with male police officials, especially in situations when they seemed patronizing toward me. By genuinely appreciating their help and regularly thanking them, they became more genuinely interested in my work – and more helpful. Police offers also proved to be most interesting people to learn from as many of them began their careers in the early 1990s when "decent police[people] were either killed or imprisoned."[4] Like Hervouet (2019), I cast a wide net and avoided narrow questions when interacting with police offices, which resulted in my collecting a wealth of data that was not always clearly related to my research topics. While Hervouet relied on his masculinity to facilitate the rapport with his respondents, as a woman, I was certainly not in a position to establish the same type of connection. However, I was subjected to particular explanations, recommendations, and guidance through mentorship and the sharing of experiences (see Turlubekova 2020), and so, unlike Sultana (2007), who felt mistreated by her male respondents, with the exception of a few disturbing incidents, I generally felt very comfortable during and after the interviews with my participants. My male participants also seemed to enjoy answering my "naive questions" and demonstrating how much more they knew about drugs than I did (Turlubekova 2020). As a matter of fact, many of my respondents were extremely polite, informed, and even charming – in complete contrast to the policemen who abused human rights or were brutal Escobar-like traffickers that I had imagined. This was clearly shown when I met a group of detectives for the first time. When I entered the room, two men were wrestling. As I hesitated to size up the situation, their supervisor sharply yelled at them and asked them to behave while "the lady" was around. I felt surprised but then smiled and said: "I think you have mistaken me for someone else. I am not a lady – I am Zhaniya. Nice to meet you." With this comment, the tension eased as everyone immediately started laughing.

It must be noted, though, that one incident did raise greater ethical considerations, which I believe happened as a result of my positionality as a young female scholar. In this situation, a respondent had suggested that he could organize my access to a wholesale heroin dealer, boasting that he had many contacts in the industry and would be happy to introduce me to them if we continued the interview in private. I still remember my goose bumps, as this was a one-on-one interview with a drug trafficker affiliated with the Russian mafia, which had begun quite late in the evening. Fortunately, there were other people in the area, but I still felt very alone with him in the office of our mutual acquaintance. My discomfort escalated as the respondent began describing his proposal even more explicitly, possibly because he saw no immediate reaction from me as my natural response was to freeze. A few seconds later, adrenaline kicked in and I came up with a quick response to remove myself from the situation; I calmly explained that while a personal meeting with a wholesale heroin dealer would be of interest, I had no interest in identity-related information or any other information that could incriminate or harm anyone. I also added that such a behavior could be recognized as unethical and have devastating consequences for my career. My respondent accepted my explanation but kept mocking me, saying that I must have affiliations with the criminal justice system, by saying that: "[n] o problem, Miss… I meant no offence… I understand that in your business even doctors have [military] rank… All hail a new criminal case!" We continued the interview in a more formal tone, and in the end the respondent declined to give his phone number but agreed to write down mine. Later on, when I told my gatekeeper about this incident, he laughed to the point of tears, advising me to read fewer theoretical books and watch more spy thrillers so that I could understand the reality on the ground. I have not fully grasped what he meant at the time, nor can I say that I fully understand it now. I did feel however that interpretation of some field dynamics would be different by people coming from different disciplines. However, soon afterwards I was distracted by arranging another interview and shifted the focus from their diverse suggestions to more important data that I had collected from him because there were still some parts I hoped to gather in the future.

A few years later while watching the Homeland series, I realized that the incident might have reminded my gatekeeper of a stereotypical "honey trap" allegedly used by the intelligence services for recruiting high-value targets. Ironically, the assumption of my respondent that I was an undercover agent, mixed with my explanations why civilians should stick to "harmless research styles," possibly frightened my respondent who was possibly imagining himself as a high-value target. I have spent much time trying to understand what may have caused his bizarre reaction: was it a manifestation of a deeply rooted fear of spies and foreign agents so common in the former Soviet Union (Hervouet 2019) or was it skepticism that a young woman of my age can truly aspire to become a scholar? Whatever the real reason was, though, I prefer to believe that my approach prevented an unfortunate turn of events that night and enhanced my understanding of the importance of risk assessments when conducting research on highly sensitive topics, especially where understandings of legality and power can be specific, uncompromising, and unknown to outsiders.

Conclusion

There are different types of information a researcher engages with during fieldwork. While searching out precious data, one might come across other less ethically reasoned data-collection options. In relation to this, Sandberg and Copes (2013) advice making "standing decisions" about how to address ethical dilemmas before entering the field and then sticking to these choices. Following their advice, as I have already mentioned, I chose to prioritize the safety of all participants including myself. This strategy made me look foolish in the eyes of some of my participants, but I believe it proved to be right. As shown, I have realized that investigating high-risk contexts might require specialized expertise, training and, in some extreme situations, institutional support because certain information might be consequential. Furthermore, I learned firsthand that scholars need the skills to be able to adjust their research priorities thematically and methodologically while in the field to allow efficient data collection even when restricted by state policies, ethical review board requirements, or other factors. However, the circumstances often change. Thus, I encourage other scholars to continue developing context-specific knowledge and seeking additional training whenever possible. I believe that there is great value in uncovering deeper layers of context-specific knowledge, understanding its undercurrents and learning to interpret its changes. In a way, it equips scholars with the means and ideas they need to conduct efficient, safe, and impactful data collection within these high-risk and highly restricted environments.

I also hope this chapter demonstrates how my fieldwork was affected by my positionality and the ways others perceived and interacted with me. Just like other local female scholars conducting research in the same region (for example, Kudaibergenova 2019; Mamadshoeva 2019; Suyarkulova 2019), I was occasionally admired, mistreated, pushed out, placed into specific categories, and expected to behave in certain ways because I was perceived as a young, naive academic. Rather than affecting my relationships with my respondents, though, my emerging mindset as a "critical criminologist" and genuine interest in their lives helped me to better explore the consequences of unequal access to justice, opportunities, and resources. Although I had previously thought, like Du Boulay (2019), that as a local, I would be in the best position to collect and interpret my data; however, as I have shown in this chapter, our positionalities often implicate our research in more complex ways than we can anticipate.

Notes

1 Part of this research was conducted while the author was an Erasmus Mundus joint doctoral candidate in Cultural and Global Criminology at Utrecht University and Eötvös Loránd University. The author therefore gratefully acknowledges the financial support from the Joint Doctorate in Cultural and Global Criminology, the European Union Education, Audiovisual and Cultural Executive Agency, the Erasmus Mundus scheme.
2 *Ty* is an informal Russian version of you that is commonly used to address inferiors, children, or those in a close relationship.
3 *Cheka* or *Vecheka* was the predecessor of the KGB and early secret police in the USSR.
4 They used "men" in this context, but I have changed it to "people."

References

Abdirov, N. 1999. *Kontseptualnye problemy borby s narkotizmom v Respublike Kazakhstan. (Kriminologicheskoe i ugolovno-pravovoe issledovanie)*. Almaty: Baspa.

Adams, L. 1999. "The Mascot Researcher. Identity, Power, and Knowledge in Fieldwork." *Journal of Contemporary Ethnography* 28 (4): 331–363. DOI: 10.1177/089124199129023479.

Adler, P. and P. Adler. 1983. "Shifts and Oscillations in Deviant Careers: The Case of Upper-Level Drug Dealers and Smugglers." *Social Problems* 31 (2): 195–207. DOI: 10.2307/800211.

Andrieu, K. 2011. "An Unfinished Business: Transitional Justice and Democratization in Post-Soviet Russia." *International Journal of Transitional Justice* 5 (2): 198–220. DOI: 10.1093/ijtj/ijr011.

Arystanbek, A. 2020. *Trapped Between East and West: A Study of Hegemonic Femininity in Kazakhstan's Online and State Discourses*. Master's Thesis, Central European University, Hungary.

Becker, H. 1967. "Whose Side Are We On?" *Social Problems* 14 (3): 239–247. DOI: 10.2307/799147.

British Society of Criminology. 2015. *British Society of Criminology Statement of Ethics*. https://www.britsoccrim.org/documents/BSCEthics2015.pdf.

Bullington, B. and A. Block. 1990. "A Trojan Horse: Anti-Communism and the War on Drugs." *Contemporary Crises* 14 (1): 39–55. DOI: 10.1007/BF00728225.

Campbell, H. 2008. "Female Drug Smugglers on the U.S.-Mexico Border: Gender, Crime, and Empowerment." *Anthropological Quarterly* 81 (1): 233–267. DOI: 10.1353/anq.2008.0004.

Cornell, S. 2005. "Narcotics, Radicalism, and Armed Conflict in Central Asia: The Islamic Movement of Uzbekistan." *Terrorism and Political Violence* 17 (4): 619–639. DOI: 10.1080/095465591009395.

De Danieli, F. 2014. "Beyond the Drug-Terror Nexus: Drug Trafficking and State-Crime Relations in Central Asia." *International Journal of Drug Policy* 25 (6): 1235–1240. DOI: 10.1016/j.drugpo.2014.01.013.

Decker, S., S. Decker and B. Van Winkle. 1996. *Life in the Gang: Family, Friends, and Violence*. Cambridge: Cambridge University Press. DOI: 10.1017/CBO9781139174732.

Du Boulay, S. 2019. "The Moral Education of a Young Woman in Kazakhstan." *openDemocracy*, December 20. https://www.opendemocracy.net/en/odr/moral-education-young-woman-kazakhstan/.

Duhamel, L. 2010. *The KGB Campaign against Corruption in Moscow, 1982–1987*. Pittsburgh: University of Pittsburgh Press. DOI: 10.2307/j.ctvl1hpslr.

Ferrell, M. and M.S. Hamm. 1998. *Ethnography at the Edge: Crime, Deviance, and Field Research*. Boston: Northeastern University.

Fujii, L. 2015. "Five Stories of Accidental Ethnography: Turning Unplanned Moments in the Field into Data." *Qualitative Research* 15 (4): 525–539. DOI: 10.1177/1468794114548945.

Golunov, S. 2007. "Drug-Trafficking through the Russia-Kazakhstan Border: Challenge and Responses." *ACTA SLAVICA IAPONICA* 24: 24–46.

Golunov, R. and R.N. McDermott. 2005. "Border Security in Kazakhstan: Threats, Policies and Future Challenges." *Journal of Slavic Military Studies* 18 (1): 31–58. DOI: 10.1080/13518040590914127.

Hervouet, R. 2019. "A Political Ethnography of Rural Communities under an Authoritarian Regime: The Case of Belarus." *Bulletin of Sociological Methodology/Bulletin de Méthodologie Sociologique* 141 (1): 85–112. DOI: 10.1177/0759106318812790.

Hunt, D. 1990. "Drugs and Consensual Crimes: Drug Dealing and Prostitution." *Crime and Justice* 13: 159–202. DOI: 10.1086/449175.

Johnson, J. 2009. "Unwilling Participant Observation among Russian *Siloviki* and the Good-Enough Field Researcher." *PS: Political Science & Politics* 42 (2): 321–324. DOI: 10.1017/S1049096509090647.

KazInform. 2020. "Glava gosudarstva podpisal popravki o protivodejstvii korrupcii." *inform.kz*, October 7. https://www.inform.kz/ru/prezident-podpisal-zakon-o-protivodeystvii-korrupcii_a3703147.

Khazanov, A. 2008. "Whom to Mourn and Whom to Forget? (Re)constructing Collective Memory in Contemporary Russia." *Totalitarian Movements and Political Religions* 9 (2–3): 293–310. DOI: 10.1080/14690760802094917.

Kudaibergenova, D. 2018. "Project 'Kelin': Marriage, Women and Re-Traditionalization in Post-Soviet Kazakhstan." In *Women of Asia: Globalization, Development, and Social Change*, edited by M. Najafizadeh and L. Lindsay, 379–390. London: Routledge.

Kudaibergenova, D. 2019. "When Your Field Is Also Your Home: Introducing Feminist Subjectivities in Central Asia." *openDemocracy*, October 7. https://www.opendemocracy.net/en/odr/when-your-field-also-your-home-introducing-feminist-subjectivities-central-asia/.

Lalander, P. 2003. *Hooked on Heroin: Drugs and Drifters in a Globalized World*. New York: Berg.

Lopega, D. 2019. "On President Rodrigo Duterte's 'War on Drugs:' Its Impact on Philippine-China Relations." *Contemporary Chinese Political Economy and Strategic Relations* 5 (1): 137–170.

Malyutina, D. 2014. "Reflections on Positionality from a Russian Woman Interviewing Russian-Speaking Women." *Sociological Research Online* 19 (4): 122–134. DOI: 10.5153/sro.3475.

Mamadshoeva, D. 2019. "Listening to Women's Stories: The Ambivalent Role of Feminist Research in Central Asia." *openDemocracy*, October 9. https://www.opendemocracy.net/en/odr/listening-to-womens-stories-the-ambivalent-role-of-feminist-research-in-central-asia/.

McCarthy, L. 2016. *Trafficking Justice*. Ithaca: Cornell University Press. DOI: 10.7591/9781501701375.

Mercille, J. 2011. "Violent Narco-Cartels or US Hegemony? The Political Economy of the 'War on Drugs' in Mexico." *Third World Quarterly* 32 (9): 1637–1653. DOI: 10.1080/01436597.2011.619881.

Milligan, L. 2016. "Insider-Outsider-Inbetweener? Researcher Positioning, Participative Methods and Cross-Cultural Educational Research." *Compare: A Journal of Comparative and International Education* 46 (2): 235–250. DOI: 10.1080/03057925.2014.928510.

Mohapatra, N. 2007. "Political and Security Challenges in Central Asia: The Drug Trafficking Dimension." *International Studies* 44 (2): 157–174. DOI: 10.1177/002088170704400205.

Noh, J. 2019. "Negotiating Positions through Reflexivity in International Fieldwork." *International Social Work* 62 (1): 330–336. DOI: 10.1177/0020872817725140.

Omelicheva, M. and L. Markowitz. 2019. *Webs of Corruption: Trafficking and Terrorism in Central Asia*. New York: Columbia University Press.

Paoli, L. 2001. "Drug Trafficking in Russia: A Form of Organized Crime?" *Journal of Drug Issues* 31 (4): 1007–1037. DOI: 10.1177/002204260103100411.

Paoli, P. and P. Reuter. 2008. "Drug Trafficking and Ethnic Minorities in Western Europe." *European Journal of Criminology* 5 (1): 13–37. DOI: 10.1177/1477370807084223.

Paoli, L., V. Greenfield and P. Reuter. 2012. "Change Is Possible: The History of the International Drug Control Regime and Implications for Future Policymaking." *Substance Use & Misuse* 47 (8–9): 923–935. DOI: 10.3109/10826084.2012.663592.

Patten, D. 2016. "The Mass Incarceration of Nations and the Global War on Drugs: Comparing the United States' Domestic and Foreign Drug Oolicies." *Social Justice* 43 (1): 85–105.

Piacentini, L. 2013. "'Integrity, Always Integrity' Laura Piacentini Argues for the Importance of Personal and Researcher Integrity in Prison Research." *Criminal Justice Matters* 91 (1): 21. DOI: 10.1080/09627251.2013.778752.

Romero-Daza, N., M. Weeks and M. Singer. 2003. "'Nobody Gives a Damn If I Live or Die:' Violence, Drugs, and Street-Level Prostitution in Inner-City Hartford, Connecticut." *Medical Anthropology* 22 (3): 233–259. DOI: 10.1080/01459740306770.

Sandberg, S. and H. Copes. 2013. "Speaking with Ethnographers: The Challenges of Researching Drug Dealers and Offenders." *Journal of Drug Issues* 43 (2): 176–197. DOI: 10.1177/0022042612465275.

Slade, G. 2013. *Reorganizing Crime: Mafia and Anti-Mafia in Post-Soviet Georgia*. Oxford: Oxford University Press.

Sultana, F. 2007. "Reflexivity, Positionality, and Participatory Ethics: Negotiating Fieldwork Dilemmas in International Research." *ACME: An International E-Journal of Critical Geographies* 6 (3): 374–385.

Suyarkulova, M. 2019. "A View from the Margins: Alienation and Accountability in Central Asian Studies." *openDemocracy*, October 10. https://www.opendemocracy.net/en/odr/view-margins-alienation-and-accountability-central-asian-studies/.

Sykes, G. 1974. "The Rise of Critical Criminology." *The Journal of Criminal Law and Criminology* 65 (2): 206–213. DOI: 10.2307/1142539.

Thibault, H. 2021. "'Are You Married?': Gender and Faith in Political Ethnographic Research." *Journal of Contemporary Ethnography* 50 (3): 395–416. DOI: 10.1177/0891241620986852.

Till, K. 2001. "Returning Home and to the Field." *Geographical Review* 91 (1–2): 46–56. DOI: 10.1111/j.1931-0846.2001.tb00457.x.

Tinasti, K. and I. Barbosa. 2017. "The Influence of Global Players on the Drug Control System: An Analysis of the Role of the Russian Federation." *Drugs and Alcohol Today* 17 (2): 124–134. DOI: 10.1108/DAT-12-2016-0031.

Turlubekova, Z. 2020. "The Devil Is in the Details: Drug-Trafficking and Corruption in Kazakhstan." PhD Dissertation, Utrecht University.

Utyasheva, L., R. Elliott, H. Canadian and A. Network. 2009. "Effects of UN and Russian Influence on Drug Policy in Central Asia." In *At What Cost? HIV and Human Rights Consequences of the Global 'War on Drugs*, edited by D. Wolfe and R. Saucier, 78–110. New York: Open Society Institute.

Varese, F. 2001. *The Russian Mafia: Private Protection in a New Market Economy.* Oxford: Oxford University Press.

Varese, F. 2017. *Mafia Life: Love, Death, and Money at the Heart of Organized Crime.* London: Profile Books Ltd.

Volkov, V. 2016. *Violent Entrepreneurs: The Use of Force in the Making of Russian Capitalism*. Ithaca: Cornell University Press. DOI: 10.7591/9781501703294.

Weitz, R. 2006. "Averting a New Great Game in Central Asia." *Washington Quarterly* 29 (3): 155–167. DOI: 10.1162/wash.2006.29.3.155.

Young, A., A. Boyd and A. Hubbell. 2000. "Prostitution, Drug Use, and Coping with Psychological Distress." *Journal of Drug Issues* 30 (4): 789–800. DOI: 10.1177/002204260003000407.

Zaitch, D. 2002. *Trafficking Cocaine: Colombian Drug Entrepreneurs in the Netherlands.* The Hague: Kluwer Law International.

Part II

Stories from the Hybrid Field

4 "Hanging Out" with the Boys

The Female Participant Observer in a Male-Dominated Group

Abigail Karas
Oxford University

Introduction

Anthropology as a discipline has long been defined by ethnography; put simply, the study of people in their own environment (Wedeen 2010, 257). Indeed, Clifford Geertz goes so far as to state that "in anthropology, or anyway social anthropology, what the practitioners do is ethnography" (Geertz 1973, 5). According to Geertz, the essence of anthropology is that it aims to be a "thick description" (1973, 6), and, to achieve this, the researcher must observe the material and social world as richly as possible. One way to reach this "thick description" is through embedding oneself within a community to develop an interpretive understanding of the social and material world. Participant observation provides the researcher with a means of achieving this understanding through long-term, sustained engagement: if we want to understand other people and their relationship to their environment, we must do so by participating in their activities and observing their relationships with the world around them.

Participant observation is not a single method of collecting data but, rather, is one comprised of several research methods. It is open-ended and, as a method of gaining insight into lived experience, is described as a prolonged period of "deep hanging out" (Clifford 1997, 188). It has been a research method at the heart of ethnographic tradition since Bronisław Malinowski wrote his seminal ethnographic work *Argonauts of the Western Pacific* in 1922 (Malinowski 1978).

Schensul, Schensul, and LeCompte define participant observation as "the process of learning through exposure to or involvement in the day-to-day or routine activities of participants in the researcher setting" (2013, 91). Tim Ingold underscored this learning by asserting that participant observation is the "ontological commitment" of anthropology, becoming a "practice of education" rather than a purely descriptive process (Ingold 2014, 388). He emphasizes that "[t]o practice participant observation, then, is to join in correspondence with those with whom we learn or among whom we study" (2014, 390). It is through this correspondence that we can make sense of the world; namely, what people say, what they do *not* say, and what is taken for

DOI: 10.4324/9781003144168-7

nted. This process of engagement with ordinary people and their day-ay encounters allows us, as researchers, to develop the "thick descrip-tion" described by Geertz. Undergoing participant observation, the researcher must therefore explore their experiences in relation to both themselves and the community being observed. From my personal experiences in the field, then, as outlined in this chapter, I hope to show what it means to be both an "emotionally engaged participant and coolly dispassionate observer" (Tedlock 1991, 69). I believe the following pages provide insights into this complex relationship, including also what it means for a female researcher studying an almost exclusively male practice.

Participation as a Method for Urban Exploration

Since 2017, my research has focused on the urban exploration practice of *roofing*, a practice that sees its participants recreationally trespass on a city's rooftops. Urban exploration (known commonly as *urbex*) is an increasingly popular activity all over the world and is defined more precisely as "the exploration of TOADS (Temporary, Obsolete, Abandoned and Derelict Spaces)" (Paiva 2008, 9). Roofing's practitioners temporarily appropriate the spaces of the city's rooftops, gaining access through courtyards and stairwells and, in extreme cases, freeclimbing. Over the past two decades, urban exploration has spread from its birthplace in North America to Russia, particularly taking root in Saint Petersburg where roofing has come to play a large part in the city's youth cultural scene. As such, Russia's northern capital and second city became the focus of my doctoral research in an attempt to contextualize the activity and explore why roofing has proven so popular in Saint Petersburg, perhaps more so than in any other Russian city.

While roofing has become an increasingly prominent activity within Saint Petersburg youth culture over the past decade, Article 139 of the Russian Criminal Code criminalizes the activity as a form of trespass. Roofing is therefore *de jure* prohibited, but the *de facto* standard is that it is shown a high degree of tolerance by both the public and policy-makers. Consequently, Russian roofers are at little risk of prosecution.

Indeed, roofing has gained such popularity in Saint Petersburg that the legalization of the activity has been discussed by municipal authorities, and illegal rooftop tours have become a popular tourist attraction, to such an extent that it can be marketed as one of the city's foremost touristic experiences (Kizyma 2020). Such tours have become so popular that they feature in the *Lonely Planet* guide to the city, and *The Calvert Journal's* 2017 *New East Travel Guide* stated that a foray onto the city's rooftops was "enough to justify a visit" to the city (Calvert Journal 2017; The Rooftop Tour in St. Petersburg 2022).

Within the Russian context, roofing is not considered transgressive in the same way as other activities taking place in public space, for example, political protests and mobilizations. In many ways, it is instead more closely aligned with mainstream youth culture in the late Soviet period. It is thus more comparable with activities considered mildly deviant, such as wearing jeans or listening to Western music, rather than to other forbidden practices which historically attracted much greater legal censure, such as drug-taking.

Although the risks inherent in rooftop navigation, and of engaging in unlawful activity, precluded me from participating in the activity of roofing itself, given the ambiguous legal status of the practice, I needed to find a methodology for my doctoral research that allowed me to gain my participants' trust both through repeated and sustained interaction in a way that was as unobtrusive as possible. While my research employed several physical and digital methods, including archival work and more structured interviews, participant observation was chosen early into my doctoral program as my primary means of data collection. When determining my methodology, I drew on the scholarship of researchers who worked on "deviant subcultures" such as gang culture (Decker and Van Winkle 1996) and drug dealers (Adler and Adler 1983). These studies make a clear case for long-term participation as the only method capable of gaining enough access to communities engaged in *de jure* illegal activities through the formation of trust because of the reality that more formal methods are likely to put off participants.

Qualitative methods were chosen in the design of my doctoral study as they "are designed to capture social life as participants experience it, rather than in categories predetermined by the researcher [as in quantitative research]" (Schutt 2006, 17). Qualitative methods thus allow for studies to develop through the fieldwork process, which means that a method like participant observation necessitates an open-ended and flexible approach. This theoretical approach is what Glaser and Strauss term *Grounded Theory*: the construction of theories through the iterative process of gathering and analyzing data (Glaser and Strauss 1967). Such an approach goes hand in hand with participant observation, allowing for constant adjustments to be made over the course of data collection according to the insights gained in the field and the key issues that emerge over time. Through the continued observation of participants, the researcher can thus gain insight by observing what they do, in contrast to what they may say they do. This methodological approach therefore reveals consistency, and indeed also inconsistency, between participants' accounts and their actions.

As an ethnographer, there are four positions one can occupy in the field: complete participant, participant as observer, observer as participant, and complete observer (Gold 1958). In many ways, participant observation requires a researcher to take on all four, moving between each in a continuum and routinely renegotiating roles with the community being studied. Whilst the researcher may start as a complete outsider, the longer time spent in the field, the stronger the relationship with participants becomes. There is, however, a paradox at the core of this methodology, as full participation tached observation may never simultaneously be achieved. Still, as rticipant and observer, data collected by these means cannot be dis- d from the researcher's own experience. It is therefore important for researchers to be aware of their own experiences within the communities being studied and how their subjectivities and biases may affect their understanding as both participant and observer (Tedlock 1991).

The relationship between illegality and ethics is an issue debated across disciplines by geographers, anthropologists, sociologists, and criminologists, among others (Dekeyser and Garrett 2021; Hamm and Ferrell 1998). While the study of illegal or dangerous practices, such as roofing, is important for better understanding the relationships people have with and within their local environments, it also poses many ethical issues. For instance, Bradley Garrett's (2013) doctoral research into an urban exploration practice similar to roofing, based in London, led to charges of conspiracy to commit criminal damage (Booth 2014). Garrett's case also highlights the different statuses of comparable activities in the UK and Russia, where the activity is illegal in the former and not only tolerated but highly popular in the latter.[1]

Nevertheless, the Research Ethics Committee for my doctoral project determined that it was too much of a risk for me to go up onto the roofs myself (due to both the inherent dangers of the roof space and the illegality of the activity, particularly given the Russian context). Although participation in such activities may involve navigating this boundary between ethics and law, this chapter highlights the possibilities offered by participant observation as a methodology for such practices, while remaining within the margins of legality.

Risk and Gender Expectations

Whilst roofing is not an exclusively male activity, it is generally perceived as such. Therefore, as I became increasingly embedded within the community throughout the duration of my fieldwork, participation in the predominantly male environment forced me to confront my position as an outsider (both as a foreigner and as a female) and as an observer. Unexpectedly, my position as a female researcher meant that, in many ways, I was well positioned to study roofing.

Roofing can be classified as a high-risk behavior: its participants work with a double concept of risk when taking part in the activity because of legal prohibitions and the hazardous terrain of the roof space. All risks are culturally biased, and risky decisions concern an individual's value judgments in being socially constructed (Douglas 1990). Anthony Giddens asserts that the fulfillment of certain individuals through risk-taking situations stems from three factors: they are aware of the risk, they voluntarily expose themselves to the risk, and they are confident that they possess the skills required to handle the risk (Giddens 1991). This formula can be applied to high-risk activities such as roofing in several ways. By choosing to take part in the activity, roofers demonstrate that they are not only well informed about the associated risks but also confident that they have the necessary skills to remain in charge; they gain satisfaction from this apparent ability to gain control over a set of seemingly uncontrollable circumstances.

This perceived ability to maintain control through the exercise of skill is a key characteristic of what Stephen Lyng terms *edgework* (Lyng 2004). In a 1974 interview with Playboy Magazine, Hunter S. Thompson described what he coined "edgework" to conjure up human experiences often considered "deviant" or "anarchic," most noticeably his experimentation with hallucinogenic drugs. The term was later adopted by Lyng in his work on voluntary risk as a form of both escape and resistance. Edgework is therefore "a form of boundary negotiation – the exploration of 'edges'" (Lyng 2004, 4). These "edges" can consequently be defined in a variety of ways: "the boundary between sanity and insanity, consciousness and unconsciousness, and the most consequential and even obvious one: the line separating life and death" (Lyng 2004, 4).

For my participants in Russia, edgework is accordingly carried out through their physical engagement with the built environment, as roofing is an embodied activity that allows its practitioners to gain a new understanding of the city and themselves. It is therefore not only a technique of the body but also a *technique of the self* (Foucault 1988). Risk decisions are largely based on individual perceptions of existing hazards, which are greatly influenced by an individual's assessment of their own skills and capacities to navigate the risks involved in the navigation of these hazards. For roofers, their athletic prowess is key to this perceived ability to navigate the rooftop terrain. The development and practical application of skill is a key aspect of the experience of edgework, as those involved in the activity probe the limits of performance in an attempt to find the boundary between order and chaos.

It is important to note that edgework is distinct from gambling: rather than making an entirely chance-based decision, edgework involves the chance to apply specific skills in the negotiation of a challenging and hazardous situation, carefully calculating the odds. Edgework also differs from gambling as it is a skill-based endeavor, wherein it is possible through practice to acquire procedures which can be adjusted to help compensate for uncertainty. Roofers thus believe that there is little chance of failure as they have the capacity to cope with the material conditions of the city's rooftop terrain, no matter how challenging this may be – this was confirmed on numerous occasions when I asked my interlocutors about the dangers of their exploits yet was regularly shrugged off. In one instance, a participant told me that skateboarding was a much more dangerous activity as you are more likely to break a wrist in a skatepark than on a roof, thus completely disregarding the greater risk to life posed by the city's rooftops. This statement accordingly functioned threefold: to reassure me of both my safety and theirs; to cognitively minimize the risks of the activity, a process key to their risk-taking decisions; and crucially, as masculine posturing, shrugging off the dangers as a façade for their own masculine desires.

Whilst the concept of ideal masculinity has been problematized in relation to a number of intersections, such as race and sexuality, Robert Connell and James Messerschmidt's (2005) reformulation of the concept of hegemonic masculinity as a pattern of practice enacted by a small dominant group within a plurality of masculinities is useful when it comes to understanding expectations of masculinity within this context. Pressures to conform to hegemonic masculine ideals may be closely linked to risk-taking behavior such as that exhibited by roofers, determining how they assess their athletic abilities, particularly in relation to other men. Through participation in roofing and the subsequent presentation of these activities, my participants therefore not only established relations with one another but also maintained their status as members of their social groups by conforming to this hegemonic masculine ideal (Clements, Friedman and Healey 2002). In fact, there is little heterogeneity within the roofing community in Saint Petersburg, as evidenced by the core group of my participants. This is because it is an activity which privileges a certain physique: young, able-bodied, athletic, and male – these attributes and their associated form of masculinity are ascribed a positive value. Moreover, the particular discourse surrounding masculinity among my roofers is also closely bound up with their willingness to engage in risk-taking.

The risk element of the practice tends to be emphasized, even fetishized, in both mainstream media and roofers' depictions of their activities: the most exposure is given to the most extreme form of the activity – those whose exploits are most daring. As a high-risk activity and also an endeavor that requires a certain degree of athleticism, roofing also induces a degree of competition between those involved. This competition manifests itself among the roofers in several ways: firstly, they wish to be the first to discover some unknown corner of Saint Petersburg; secondly, they strive to climb the highest and to take the biggest risks; and thirdly, there is competition to receive the most likes and engagement on social media (a motivating factor that is closely linked to the first two).

As risk-taking is expected of men in a way that it is not of women, especially within the Russian context, and since there is evidence that suggests men also perceive less inherent risk in their actions (Weber, Blais and Betz 2002; Baker and Maner 2008, 2009; Wang, Kruger and Wilke 2009), this gendered expectation of risk was central to my ability to carry out my research with this group without endangering my own safety. In particular, as a female, there was never the expectation for me to match the bravado of my participants or to take the same risks that they take on a daily basis.

Changing Positions and Saving Face

I was involved to some degree or another with the roofing comm Saint Petersburg since 2017, although the most substantial part of work took place between August 2019 and March 2020, when it wa to an abrupt halt by the COVID-19 pandemic. Initially, my participants viewed me with a sense of curiosity – they could not understand why I would possibly want to study them or roofing but were gracious enough to let me "hang out." Indeed, in the first days of my fieldwork, they were more eager to ask me questions about what I was doing than to talk about roofing at all. This attitude was greatly influenced by their perceptions of academia as being stuffy, old-fashioned, and the purview of old men (all my participants had completed high school, but only one had continued into higher education). Hence, in their eyes, I did not fit the stereotype of an academic. I had been introduced to them through one of their girlfriends at the time – a friend I had made as an undergraduate on my year abroad – and so was seen as very nonthreatening. Although my physical stature (I stand 4 feet 10 inches) was a constant reminder of my outsider position as a woman in a male-dominated group, in many ways, this benefitted me in the initial stages of my fieldwork as my presence was not seen as hostile or dangerous but, on the contrary, I was seen as soft and harmless.

As the gatekeeper to the group was the boyfriend of my friend (a tangential relationship and I did not know him at all well before initially embarking on my research), I was able to make connections through him as he vouched for me. This also meant that I quickly established a purely platonic relationship with him, which acted as a benchmark for my relationships with the other members of the group. This allowed me, for the most part, to avoid sexualized or demeaning comments. I was also able to deflect flirty jokes or comments early on, making it clear that I would not put up with such remarks whilst also laughing them off or rolling my eyes. It was, nonetheless, very clear early on that these were only lighthearted jokes with no serious intent and that they were no different from the comments and jokes I might receive from friends at home in the UK – they quickly learned I was not going to engage with any sort of flirty or romantic behavior. Still, it was a fine line between showing that I could "take a joke" and setting boundaries for myself with the boys, and therefore, it proved important that I spoke their language and could engage in their banter in order to diffuse any awkwardness. Whilst some researchers have been subjected to demeaning comments from participants (Lumsden 2009; Haddow 2021), if anything, these interactions not only boosted my position within the group in showing that I could engage in some harmless jokes but also established clear boundaries regarding potential romantic relationships.

I further found that adopting an almost genderless persona was helpful for my research, so jeans and a sweatshirt became my fieldwork uniform –

a tactic that has been adopted by other researchers with varying degrees of success (Maher 1997; Bucerius 2013). In Russia, heteronormativity is politically and socially enforced, and citizens are clearly polarized along gendered lines. Just as hegemonic masculinity is considered the most prestigious form of being a "man," a particular style of femininity is privileged to reinforce hegemonic beauty ideals. In many contexts, it may indeed be more appropriate to perform the privileged form of femininity; for example, during her own experience of working with young men, Sandra Bucerius (2013) found that her attempts to minimize outward signifiers of gender drew increased negative comments from the communities she was researching. Yet, within the social group I was studying, it was more effective for me to adopt a midway stance whereby I did not draw as much attention to my gender as perhaps would be appropriate within other settings. That is not to say that it was not present at all – my short stature and long hair were reminders – but I chose to adopt a less-emphasized form of femininity than perhaps is expected in Russia due to traditional gender norms and practices. I strongly believed that by styling my gender in this way, I was able to avoid potentially awkward or uncomfortable interactions with not only my participants but also their wider social circles in which I was embedded.

Accounts from female ethnographers working within male-dominated environments also documented sexist behavior and, in particular, "hustling" as a form of sexualization which can serve to demean or devalue the position of the researcher (Gurney 1985; Haddow 2021). Despite what I had read before entering the field, and the preconceived notions I had from cautionary tales told to me by more senior researchers, I was lucky that my participants were nothing but respectful to and of me throughout my field research. In retrospect, I believe that many of my reservations before spending time with the group were due to stories I had been told from a young age about being careful around men, especially strangers. I believe my proximity in age to my participants contributed to the ease of our relationship, though, as they treated me as one of their peers from the very beginning. Although I could never expect to be afforded the title of *bratan* or even *chuvak*,[2] the less attention I drew to my gender – as a key marker of my outsider status – the easier it was for me to engage with the group.

Whilst at the start of my fieldwork I felt my outsider position was constantly magnified, as time moved on my relationship with the group changed: I was able to establish a position for myself within the clique, partially integrated as *svoi* (one of our own). The first time I was referred to as "one of us" was, for me at least, a pivotal point in my fieldwork. It was around this time that invitations were more freely extended to me, and I was able to participate in and observe a much wider array of activities beyond roofing – from hanging out at someone's house to skateboarding, and even a trip to the suburbs in *Leningradskii oblast'* (an age-old Petersburg tradition).

The more integrated I became within the group, the more important it was for me to evaluate my position as an academic, toeing the line between researcher and social acquaintance (Tillmann-Healy 2003). There is a fine ethical boundary that must be navigated between being friends with participants and having a researcher-participant relationship; on the one hand, friendship can provide a unique insight into the social lives of participants. However, one of the key ethical issues of participant observation is that participants have no obligation to researchers as either academics or friends, and therefore, researchers must try to balance their projects with their duty of care to all involved. It is worth underscoring that even a researcher whose demographics closely align with their participants needs to constantly appraise their relationship to the group, as reflexivity is central to the methodology, although this might be more difficult when these differences are not so acutely magnified.

It is, of course, important to note that in many social settings which were a part of my participants' lives, outside of their roofing activities, other women were often present. Several of my participants also live in female-dominated households with their mothers, sisters, and, occasionally, also grandmothers. Although it was useful to talk to the other women about their attitudes toward roofing, I felt it was important that I was not pressured to join them socially, rather than the core group of my participants, on account of our shared gender. This was especially important to me given that I had initially been introduced to the group by one of their girlfriends and thus already had established relationships with the female members of the extended circle, yet, I had never been close to my participants before my research.

Of course, just as I felt I had established myself within the group where my gender was no longer a barrier to access, two of my participants bought me a plant for International Women's Day. Although the state of the plant suggested it was hurriedly purchased from the nearest shop on the way to meet me, this thoughtful gesture served to remind me of my position within, or rather, on the fringes of, the group. I have since tried to think whether there might be an equivalent situation in which a male researcher might find themself, but none come to mind. This is particularly the case because, whilst Defender of the Fatherland Day, celebrated on February 23, is often regarded as the male equivalent of Women's Day – and women may equally give small gifts to the men in their lives on this day – there is no comparison in terms of the scale of celebration and the relative cultural impacts of the two holidays (Kozlova 2014). Indeed, International Women's Day may therefore be seen as an expression of the lack of gender parity in contemporary Russia and, perhaps, the post-Soviet space more largely.

Thus, as a female researcher attempting to establish a place for herself within a predominantly male social group, setting boundaries for my participation was crucial. Although roofing is granted a high degree of social tolerance, as I noted, it is still, in essence, an illegal activity. Trust between myself and my participants was therefore invaluable, and so establishing the degree to which I was to participate and observe was crucial in the early stages of my study. Although institutions are becoming increasingly open to the need for a "situated ethical understanding" (Dekeyser and Garrett 2021), in my case, I was precluded from participation in the activity of roofing itself, given its illegality and high-risk nature. This was despite acknowledgement by the anonymous reviewers from my Research Ethics Committee that the illegality of the activity was "modest" and, also that I was not encouraging my participants in their risk-taking (it was an activity in which they were already engaged).

For some researchers, this judgment may have been extremely problematic, threatening their status within the group. Yet, as there was already no expectation for me to participate in every aspect of their activities due to my prior experiences with them, I was able to continue without losing trust or face within the group. As David Yau-fai Ho describes: "losing face is a serious matter which will, in varying degrees, affect one's ability to function effectively in society" (Yau-fai Ho 1976, 867, cf. Goffman 1955). Although initially, I believed that not being able to access the roofs would be a major pitfall in my research, it did encourage me to reappraise the focus of my work; instead of looking at the embodied practice of roofing, I became more concerned with the wider context in which the practice is located. I began to explore the broader cultural resonance of the roofs in the local imagination. Saint Petersburg, as opposed to other cities in Russia, had been selected as the research site due to the popularity of roofing, and so, it was not difficult for me to expand the topic to accommodate this obstacle, increasing the scope of my topic to include a much wider array of social groups and uses of the rooftops: from literary and artistic depictions of the Saint Petersburg skyline, which have cemented the iconic nature of the roofs in the local imagination, to the proliferation of rooftop bars and restaurants.

Indeed, it was in these ground-level spaces that I began to understand how the social relations which were initially facilitated by the activity, and a shared identity of "roofer," were maintained away from the rooftop space. For the most part, roofing is practiced either individually or in small groups (a fact necessitated by its illegality), so there are clear spatial differences between the areas in which my participants, like many roofers, live and the areas where their activities are concentrated. As it is an activity which seemingly takes place away from their normal lives, it was easy for me to act as a participant-observer away from the roof space by taking part in the other activities that make up their daily lives.

In their work on Russia's skinhead movement, Pilkington, Omel'chenko, and Garifzianova (2013) argue for a shift away from "subculture" as an analytic concept, indicating a more fluid relationship between "subcultural life" and "everyday life." They contend that the so-called subcultural traits shared by the group – this may be a shared style, ideology, or in the case of roofing, an activity – constitute only one part of their day-to-day lives which also include work and family (Pilkington, Omel'chenko and Garifzianova 2013). As such, when considering the demographics of roofers, it is also interesting to consider how the practice may fit into the rest of their biography. My participants had a number of jobs throughout my time in the field, ranging from retail to working in the city's *Baltika* brewery. Although I was unable to follow them to work, through participant observation I was able to locate roofing within the context of their everyday cultural strategies. Normally, people do not put their experiences of the world into words; they take their routinized, habitual experience for granted. Thus, hanging out in a variety of settings and participating in these everyday experiences allows a researcher to gain some understanding of the routinized behavior of their participants. In this way, it is possible to share a mutual understanding of the world with one's participants, which is vital for gaining insight into the wider structural circumstances underpinning certain customs or conventions and, in my case, have allowed an activity such as roofing to flourish.

For these reasons, my participation extended to a range of everyday social settings where my role of participant-observer was magnified. On several occasions, being a female researcher allowed me to preserve my role and social standing in situations that might have been more challenging had I been a male researcher. For example, there was no expectation for me to participate in drinking as there may have been otherwise – it was not offensive for me to decline a beer if I was invited to a *tusovka* (hang out). I could freely do so without risking my status within the group or my next invitation to hang out. It became a joke that I would always bring a bottle of Diet Coke to a gathering. This may have marked me as a foreigner but, luckily, seemed to be without any detriment to my social standing and more a point of amusement. I believe that this spoke to the generation of my participants, who were all born in the 1990s and grew up in the immediate post-Soviet period. Thus, they had experienced Western brands throughout their lives and were much more accepting of this little quirk than perhaps a Russian of an older generation might have been. It is also worth highlighting the impact that researching Saint Petersburg as opposed to any other Russian city may have had, specifically the openness of my participants to foreigners was likely greatly impacted by the city, which was founded by Peter I to be Russia's "Window to the West" and still prides itself on its cosmopolitan identity.

One of the primary benefits of participant observation is that it is a comparatively unobtrusive and natural method of data collection and thus possible to rely not only on verbal but also nonverbal cues. In my experience, I was therefore able to contrast how my participants spoke about their activities with both their presentation of roofing on social media and their real-life actions. In group conversations, I had to decide to what extent I wanted

:k and watch my participants interact, taking mental notes of every .etail of the group dynamic, or take part in the conversations in order .ɔ ask questions as a very relaxed mode of interviewing. It was in these informal contexts that rapport between myself and my participants was built as we chatted and got to know one another, but I was also able to observe intergroup dynamics and gain contextual information about the roofers and their lives, which might otherwise be taken for granted. Knowing when to participate and when to observe thus proved crucial, revealing information that was not evident on first meeting. The flexibility and hybridity of participant observation as a methodology allowed me to respond naturally to each individual conversation.

Moreover, it was in these situations that I was able to gain respect and, if anything, boost my status within the group as they did not take themselves too seriously and prided themselves on their witty back-and-forth banter. As it would be uncomfortable for everyone, myself included, had I silently sat in the corner with a notebook watching these exchanges, my participation was required as part of my observation. Indeed, I believe that had I opted to be a more passive bystander, I would have missed some of the most interesting and intimate conversations. In this way, I had to navigate the discrepancies between being an academic and being a participant carefully and elected to take part, as I would in many other social situations; I would occasionally take notes on my phone if I thought something was extremely important or if someone had come out with a memorable quote that had to be included verbatim. Still, I generally waited until I had time to hurriedly jot things down in order to avoid interrupting the flow of conversation. This is a difficult task but, like anything, became easier with practice and more time spent in the field. The long-term format of participant observation is therefore important as I could follow up on things that I had heard in conversations after the fact with more structured interviews, whether for clarification or to explore an issue in greater depth.

Yet, even in these settings, it still proved important for me to realize when to take a step back and observe my participants. During my fieldwork, I was present during several highly uncomfortable conversations involving racist and xenophobic remarks by my participants. This was particularly prevalent in the first few months of 2020 when the escalating COVID-19 pandemic fueled increased Sinophobia among my participants. Such views are not particularly uncommon or shocking in the Russian context, yet I was still forced to reflect regularly on our social and cultural differences on these occasions. Nonetheless, these comments, at times, made me feel extremely uncomfortable, more so than any aspect of my gender did throughout the entire research project.

Whilst in a regular social situation at home in the UK, I would have felt compelled to counter these opinions, I had to remember my role and consider that I may potentially alienate my participants and jeopardize my position within the group. Our different cultural upbringings were also magnified in these conversations, and I did not want to draw attention to my outsider status, so staying silent allowed me to save face. Since relationships are always multifaceted, I found it complex to navigate my feelings of friendship toward my participants when they espoused these troubling views. Nevertheless, it was through the research process, by taking on the role of researcher, participant, observer, and also friend, that I learnt to be more open-minded and understanding (Tillmann-Healy 2003). Whilst such incidents muddied the boundaries between a participant and observer, they also forced me to evaluate how I interact with my participants and those with views which oppose my own. My experiences therefore showed me the value of compassion and patience during the ethnographic process.

My fieldwork experiences taught me the necessity of reflexivity: it was an experience of learning, not only about the community being studied but also about myself as a researcher and as an individual. Thus, being mindful of one's subjectivity during the research process is important, especially as a participant-observer, as it involves studying not only a community but, in many ways, also oneself and one's social relations with their participants.

By embedding myself within a community that exists in the social and material world, I realized that we as researchers inherently become a part of that which we are studying. At times, this may seem paradoxical; as Benjamin Paul writes: "[p]articipation implies emotional involvement; observation requires detachment. It is a strain to try to sympathize with others and at the same strive for scientific objectivity" (1953, 441). Whilst I read this passage prior to entering the field, it was not until I was faced with the challenge of navigating my place within a social group that I truly understood the difficulty described by Paul. I would, however, argue that this emotional involvement can be a benefit to us as researchers as long as we take into consideration that our experiences will always be based on our own implicit biases. Our subjectivities are useful as long as they motivate us to think critically about our impact on the research and the data collected, as well as the community under study. Indeed, positionality is a key consideration given that we are "observing through participating such that the self becomes the primary research tool" (Evans 2012, 98). Accordingly, then, participant observation requires a conscious understanding not only of the community being studied but, crucially, also of oneself.

Conclusion: Participant Observation during a Pandemic

As outlined, participant observation is a methodology that requires sustained immersion in a field in order to gain a shared cultural understanding of participants within their own settings. Following from that, what does a participant observer do when there is a global pandemic and, suddenly, the social interactions forming the basis of their research are curtailed? In March 2020, I was forced to suspend my fieldwork and return to the UK from Russia as coronavirus cases rose sharply and lockdowns loomed in both countries.

Participation, evidently, was a core element of my research prior to the pandemic; it helped me to make sense of the norms and feelings governing my participants' actions. In the case of roofing, participant observation while I was physically in Russia was initially the only way for me to collect data. Given the nature of roofing as an activity, laying a foundation of trust was accordingly key, as it required me to be physically present with my participants. But whilst it yielded insightful results, the COVID-19 pandemic helped me to realize that participant observation does, in fact, function best when used alongside other methods. I therefore substantiated the data collected through participant observation with archival work and more structured interviews with both my participants and other residents of Saint Petersburg, including artists, photographers, and architectural preservation activists. Even when confronted with several lockdowns, I was able to employ digital methods to maintain momentum in the hope that I would be able to return to the field.

In a pre-pandemic world, my participants were reluctant to talk to me digitally and so spatial and temporal proximity was essential if I wanted to elicit more than a monosyllabic "yes/no" answer to my questions (Jones 1999; Miller and Slater 2000; Murthy 2008). This is fairly typical when conducting research in Russia, where meeting face-to-face is an important step in forming trust along with vouching, as is evident in studies of informal practices (e.g., Ledeneva 1998). It is thus particularly important when studying an activity which, like roofing, has a degree of criminal responsibility and was why my access to the group via a friend was so important. Within this context, the relative long-term format and physical propinquity of participant observation are what make it such an effective method. Although my presence in the field was essential to understanding social dynamics and individual motives, my participants became more willing to talk to me virtually than ever before (partly out of boredom) during the pandemic. This is an experience that many researchers globally have observed since the global lockdowns of 2020 (e.g., Howlett 2021).

Whilst online ethnographic fieldwork was insufficient in the first instance, the fact that I had already developed an understanding of the roofing community allowed me to maintain these relationships during the first lockdown and enabled me to continue much of my research through the pandemic. However, using online tools for data collection did also complicate my relationship with my participants. In earlier stages of my fieldwork, there were clear lines drawn every time I entered and left the field: I would communicate with them when I was due to be in Saint Petersburg, but we communicated less when I returned home to the UK. Now, it was much more difficult to compartmentalize my work, and I felt a greater duty to continuously stay in touch. This required more emotional labor from me, as the researcher, which was tough given the added pressures and anxieties of the pandemic (Knott 2019).

It is therefore worth noting that digital ethnography raises a new set of complex ethical issues beyond those conventionally raised by participant observation (Ebo 1998). Before the COVID-19 pandemic, my research included some digital methods, as I looked at the online presentation of the activity and attitudes toward my participants. However, during the pandemic, I was suddenly forced to continue my participant observation within the digital realm, interacting with my participants just as they were interacting with each other due to the lockdown. As people have come to weave together their lives on- and offline as a result of new and diverse communicative platforms, what Mirca Madianou and Daniel Miller term "polymedia" (Madianou and Miller 2011), new digital environments have emerged to mediate these relationships which exist both digitally and in the real world. Given that people are now spending much more time online, these new mediated forms of interaction only further contextualized those in which I had previously participated and observed, consequently pushing me to reappraise the spaces and practices of roofing that I study, and indeed my participants' lives and even my own, with both physical and digital realms.

Notes

1 In fact, almost anyone you speak to in Saint Petersburg will at least be aware of roofers, and young locals have often participated in some form or another of roofing themselves, even if it is just "hanging out" on a rooftop.

2 The English equivalents would be something like "bro" and "dude," but without the perhaps outdated connotations.

References

Adler, P. and P. Adler. 1983. "Shifts and Oscillations in Deviant Careers: The Case of Upper-Level Drug Dealers and Smugglers." *Social Problems* 31 (2): 195–207. DOI: 10.2307/800211.

Baker, M. and J. Maner. 2008. "Risk-Taking as a Situationally Sensitive Male Mating Strategy." *Evolution and Human Behaviour* 29 (6): 391–395. DOI: 10.2307/800211.

Baker, M. and J. Maner. 2009. "Male Risk-Taking as a Context-Sensitive Signalling Device." *Journal of Experimental Social Psychology* 45 (5): 1136–1139. DOI: 10.1016/j.jesp.2009.06.006.

Booth, R. 2014. "Oxford University Academic Who Scaled Shard Is Spared Jail Sentence." *The Guardian*, May 22. https://www.theguardian.com/education/2014/may/22/oxford-university-academic-shard-jail-place-hacker-garrett.

Bucerius, S. 2013. "Becoming a 'Trusted Outsider': Gender, Ethnicity, and Inequality in Ethnographic Research." *Journal of Contemporary Ethnography* 42 (6): 690–721. DOI: 10.1177/0891241613497747.

Calvert Journal. 2017. "New East Travel Guide". Mobile Application. 2017.
Clements, B., R. Friedman and D. Healey. 2002. *Russian Masculinities in History and Culture*. Basingstoke: Palgrave Macmillan. DOI: 10.1057/9780230501799.
Clifford, J. 1997. "Spatial Practices: Fieldwork, Travel, and the Disciplining of Anthropology." In *Anthropological Locations: Boundaries and Grounds of a Field Science*, edited by A. Gupta and J. Ferguson, 185–222. Berkeley: University of California Press.
Connell, R. and J. Messerschmidt. 2005. "Hegemonic Masculinity: Rethinking the Concept." *Gender & Society* 19 (6): 829–859. DOI: 10.1177/0891243205278639.
Decker, S. and B. Van Winkle. 1996. *Life in the Gang: Family, Friends, and Violence*. Cambridge: Cambridge University Press. DOI: 10.1017/CBO9781139174732.
Dekeyser, T. and B. Garrett. 2021. "Illegal Ethnographies: Research Ethics beyond the Law." *Research Ethics in Human Geography*, edited by S. Henn, J. Miggelbrink and K. Hörschelmann, 153–167. London: Routledge. DOI: 10.4324/9780429507366-9.
Douglas, M. 1990. "Risk as a Forensic Resource." *Daedalus* 119 (4): 1–16.
Ebo, B. 1998. *Cyberghetto or Cybertopia? Race, Class and Gender on the Internet*. London: Prawger.
Evans, G. 2012. "Practising Participant Observation: An Anthropologist's Account." *Journal of Organizational Ethnography* 1 (1): 96–106. DOI: 10.1108/20466741211220697.
Foucault, M. 1988. "Technologies of the Self." In *Technologies of the Self: A Seminar with Michel Foucault*, edited by M. Foucault, L. Martin, H. Gutman and P. Hutton, 16–49. Amherst: University of Massachusetts Press.
Garrett, B. 2013. *Explore Everything: Place-Hacking the City*. London: Verso.
Geertz, C. 1973. *The Interpretation of Cultures: Selected Essays*. New York: Basic Books.
Giddens, A. 1991. *Modernity and Self-Identity: Self and Society in the Late Modern Age*. Cambridge: Polity.
Glaser, B. and A. Strauss. 1967. *The Discovery of Grounded Theory: Strategies for Qualitative Research*. Chicago: Aldine.
Goffman, E. 1955. "On Face-Work: An Analysis of Ritual Elements in Social Interaction." *Psychiatry* 18 (3): 213–231. DOI: 10.1080/00332747.1955.11023008.
Gold, R. 1958. "Roles in Sociological Fieldwork Observations." *Social Forces* 36 (3): 217–223. DOI: 10.2307/2573808.
Gurney, J. 1985. "Not One of the Guys: The Female Researcher in a Male Dominated Setting." *Qualitative Sociology* 8 (1): 42–62. DOI: 10.1007/BF00987013.
Haddow, K. 2021. "'Lasses Are Much Easier to Get on with': The Gendered Labour of a Female Ethnographer in an All-Male Group." *Qualitative Research*: 1–15. DOI: 10.1177/1468794120984089.
Hamm, M. and J. Ferrell. 1998. *Ethnography at the Edge: Crime, Deviance, and Field Research*. Boston: Northeastern University Press.
Howlett, M. 2021. "Looking at the 'Field' through a Zoom Lens: Methodological Reflections on Conducting Online Research during a Global Pandemic." *Qualitative Research* (January): 1–16. DOI: 10.1177/1468794120985691.
Ingold, T. 2014. "That's Enough about Ethnography!" *HAU: Journal of Ethnographic Theory* 4 (1): 383–395. DOI: 10.14318/hau4.1.021.
Jones, S. 1999. *Doing Internet Research: Critical Issues and Methods for Examining the Net*. London: SAGE. DOI: 10.4135/9781452231471.

Knott, E. 2019. "Beyond the Field: Ethics after Fieldwork in Politically Dynamic Contexts." *Perspectives on Politics* 17 (1): 140–153. DOI: 10.1017/S1537592718002116.

Kozlova, N. 2014. "International Women's Day and the Construction of the Soviet Gender System." *Women's History in Russia: (Re)Establishing the Field*, edited by M. Muravyeva and N. Novikova, 137–155. Newcastle upon Tyne: Cambridge Scholars Publishing.

Kizyma, R. 2020. 'V Peterburge khotiat legalizovat' progulki po krysham'. *RBK*, 25 March 2020. https://www.rbc.ru/spb_sz/25/03/2020/5e7b1a339a7947c75c2b51fa.

Ledeneva, A. 1998. *Russia's Economy of Favours: Blat, Networking and Informal Exchange*. Cambridge: Cambridge University Press.

Lumsden, K. 2009. "Don't Ask a Woman to Do Another Woman's Job: Gendered Interactions and the Emotional Ethnographer." *Sociology* 43 (3): 497–513. DOI: 10.1177/0038038509103205.

Lyng, S. 2004. *Edgework: The Sociology of Risk-Taking*. London: Routledge. DOI: 10.4324/9780203005293.

Madianou, M. and D. Miller. 2011. *Migration and New Media*. London: Routledge. DOI: 10.4324/9780203154236.

Maher, L. 1997. *Sexed Work: Gender, Race, and Resistance in a Brooklyn Drug Market*. Oxford: Clarendon Press.

Malinowski, B. 1978. *Argonauts of the Western Pacific: An Account of Native Enterprise and Adventure in the Archipelagoes of Melanesian New Guinea*. Prospect Heights: Waveland Press.

Miller, D. and D. Slater. 2000. *The Internet: An Ethnographic Approach*. Oxford: Berg. DOI: 10.5040/9781474215701.

Murthy, D. 2008. "Digital Ethnography: An Examination of the Use of New Technologies for Social Research." *Sociology* 42 (5): 837–855. DOI: 10.1177/0038038508094565.

Paiva, T. 2008. *Night Vision: The Art of Urban Exploration*. San Francisco: Chronicle.

Paul, B. 1953. "Interview Techniques and Field Relationships." In *Anthropology Today: An Encyclopedic Inventory*, edited by A. Kroeper, 430–451. Chicago: University of Chicago Press.

Pilkington, H., E. Omel'chenko and A. Garifzianova. 2013. *Russia's Skinheads: Exploring and Rethinking Subcultural Lives*. Abingdon: Routledge.

Schensul, S., J. Schensul and M. LeCompte. 2013. *Initiating Ethnographic Research: A Mixed Methods Approach*. Lanham: Altamira Press.

Schutt, R. 2006. *Making Sense of the Social World: Methods of Investigation*. 2nd edition. London: Pine Forge Press.

Tedlock, B. 1991. "From Participant Observation to the Observation of Participation: The Emergence of Narrative Ethnography." *Journal of Anthropological Research* 47 (1): 69–94. DOI: 10.1086/jar.47.1.363058.

The Rooftop Tour in St. Petersburg. 2022. *Lonely Planet*. 2022. https://www.lonelyplanet.com/russia/st-petersburg/activities/the-rooftop-tour-in-st-petersburg/a/pa-act/v-35261P1/360547.

Tillmann-Healy, L. 2003. "Friendship as Method." *Qualitative Inquiry* 9 (5): 729–749. DOI: 10.1177/1077800403254894.

Wang, X., D. Kruger and A. Wilke. 2009. "Life History Variables and Risk-Taking Propensity." *Evolution and Human Behaviour* 30 (2): 77–84. DOI: 10.1016/j.evolhumbehav.2008.09.006.

Weber, E., A. Blais and N. Betz. 2002. "A Domain-Specific Risk-Attitude Scale: Measuring Risk Perceptions and Risk Behaviours." *Journal of Behavioural Decision Making* 15 (4): 263–290. DOI: 10.1002/bdm.414.

Wedeen, L. 2010. "Reflections on Ethnographic Work in Political Science." *Annual Review of Political Science* 13 (1): 225–272. DOI: 10.1146/annurev.polisci.11.052706.123951.

Yau-fai Ho, D. 1976. "On the Concept of Face." *American Journal of Sociology* 81 (4): 867–884. DOI: 10.1086/226145.

5 Balancing Diasporic Ties and Research: A Ukrainian-Canadian's Reflection on Fieldwork in Ukraine

Marnie Howlett[1]
University of Oxford

Introduction

> Following a formal one-hour discussion about the geography, politics, and culture of the region where we were located, I thanked the regional administrator for her time and casually mentioned how nice it was to return to a country with which I have familial ties. Upon hearing these words, she paused, smiled, and excitedly exclaimed, "[y]ou are part of the diaspora!? Come visit me when you return to the region. I'll take you to a few villages, and you will get to really experience the culture here. It makes me happy to know that you are conducting research in Ukraine, where your family is from."

By its very nature, ethnographic field research is immersive. Spending significant periods of time embedded in the sociocultural environment that one studies can yield fascinating and, very often, unpredictable results. Yet as data are collected through interactions with people and places in a particular field site, immersive approaches to research may also come with unique ethical, methodological, and even ontological challenges as a result of scholars' positionalities and subjectivities. Several key works on ethnography and interpretivist methodologies have therefore outlined the importance of reflexivity when conducting social science research (see, for example, Schatz 2009; Wood 2009; Schwartz-Shea and Yanow 2011; Yanow and Schwartz-Shea 2014), which Roni Berger outlines is "a continual internal dialogue and critical self-evaluation of the researcher's positionality" (2015, 200). Within the context of field research, Lee Ann Fujii (2016) further posits that there are many benefits of reflexivity, as engaging in explicit "self-aware meta-analysis" (Finlay 2002, 209) or the "process of self-examination…through specific actions" (Probst 2015, 38) ensures a greater cognizance of the ascribed identities that people hold – such as their visible features, genders, ages, hair, and skin colors – and the ways these attributes can, and do, influence the interactions between researchers and the participants in their studies. Other implicit and less visible characteristics may also innately define the power dynamics of various relationships in the field, such as language skills, prior knowledge, experiences in familiarity

DOI: 10.4324/9781003144168-8

with a research site, and even ancestral or personal ties to a certain place or country. By encouraging academics to gaze both outwards and inwards at his/herself in considering their subjectivities, the act of being reflexive thus incites more awareness of, and deliberation about, the ethical dimensions of research and the meanings around researcher-participant interactions in the field (Fujii 2016).

In my own case, having a diasporic connection to Ukraine proved particularly challenging to navigate while conducting fieldwork during my doctoral studies. As the opening vignette reveals, the fact that I had family ties to the country was reason enough for a political figure to change her demeanor, approach, and overall perception of me. This story is not an anomaly; I had many similar experiences during the time I spent in Ukraine as a doctoral student. Although these interactions both opened up new opportunities to explore my field sites and introduced me to diverse networks across the country, the inter-personal dynamics between my participants and myself sometimes made it incredibly difficult for me to reconcile my identities as both a researcher and member of the Ukrainian diaspora. Through my time in the field, I hence came to realize firsthand what is outlined in much of the methodological literature I had consulted prior to conducting field research; namely, that our implicit biases and subjectivities very much shape our research and the larger research process, and can easily encourage unforeseen moral and ethical quandaries. In speaking to the broader sociopolitical context of Ukraine and other post-Soviet societies, my experiences in the field accordingly highlight the importance of reflexivity in qualitative, and especially ethnographic, research. Precisely, the reflections in this chapter underscore the need for us scholars to more explicitly recognize the larger consequences of our interventions and interactions within the local sites we study to ensure more ethical and transparent research.

This chapter therefore draws on stories and observations from the four months I spent conducting field research in Ukraine prior to the COVID-19 pandemic. In reflecting on some of the methodological and ethical challenges I encountered as a result of my own subjectivities, I highlight how my diasporic connection to the country, in addition to my identity as a young female scholar, influenced the progression of my work. The following pages point to the advantages that came with being a Ukrainian-Canadian researching Ukraine, especially in terms of establishing rapport and gaining access to field sites and new networks, as well as revealing the difficulties in navigating the various identities one holds whilst ensuring transparent and ethical research. As there is currently little written about the conduct of field research in Ukraine, the former Soviet space, or even a country we may be very familiar with or have ancestral ties to, the insights included here – though focusing mainly on ethnographic field research – have relevance across methodological and disciplinary boundaries, and in other regional contexts. The chapter also makes mention of the

ways the COVID-19 pandemic forced me to alter my research desi
approach to fieldwork, and therefore also the ways that I was able to
act with my participants, which may prove insightful for how we t
about field research in a post-pandemic world. It is my hope that the re
tions in the following pages prove useful for other academics who have not (yet) gone to the field or are still in the planning stages of their research, and also elicit a larger conversation about the experiences of early-career scholars, and especially female scholars, conducting field research in the post-Soviet space.

Diaspora and Diasporic Identity

Social science scholars have increasingly used the term "diaspora" in studies of nationalism and identity (see, for example, Brubaker 2005; Safran 2005; Anthias 2008). Often employed to explain the forces and phenomena of transnational movements, it is defined by the Cambridge Dictionary (2020) as "a group of people who spread from one original country to other countries." Yet, the loose usage of the term neither critically nor accurately encompasses the diverse patterns of mobility and settlement of diaspora populations, nor the experience of dispersal, displacement, or de-territorialization in simultaneously being attached to more than one political or geographical entity (Clifford 1997). Diasporas, as William Safran (2005) explains, are unique in that they have been scattered (by choice or by forced relocation) from an original homeland or "center" to two or more host or "peripheral" regions. Though the term "diaspora" is often used interchangeably with other related designations, such as expatriates, refugees, and exile or overseas communities, diasporic populations are unique and "exemplary communities of the transnational moment" (Tölölyan 1991, 5). In this way, they are different from other migrant groups – they are neither a collective from the same country nor a group of people who are distinguished by some secondary characteristic.

James Clifford (1997) furthers that shared consciousness and experience are important and, arguably, even essential to the development and existence of diasporic populations; he notes that diasporas often struggle to define themselves as a distinct community in a new country and in light of their historical displacement (see also Clifford 1994; Anthias 2008). Whilst some scholars emphasize the linkages (real or imagined) between a homeland and a diaspora (Safran 1991), Clifford places a greater emphasis on the complex sociopolitical formations of diasporic populations themselves, explaining that they "do not come from elsewhere in the same way that 'immigrants' do" (1997, 250). Safran echoes a similar argument in outlining that diasporas commonly have a complicated and uneasy relationship with their hostland and fellow citizens because they "retain a collective memory, vision, or myth about their original homeland – its physical location, history, and achievements, and, often enough, sufferings" (2005, 37). Accordingly, diaspora communities distinguish themselves from their host society in believing they cannot be fully accepted or integrated into the larger society (Safran 2005), consequently becoming a "separate society or quasi-society in a larger polity" (Armstrong 1976, 394). Rogers Brubaker (2005) further outlines that a homeland orientation is an important element in the constitution of a diaspora. In particular, diaspora communities continue to relate to their homelands, whether personally or vicariously, when constructing themselves as collectives, and often define their ethno-communal consciousness and solidarity in terms of its safety, independence, and prosperity (Safran 2005). Despite this direct connection to a larger homeland, it must be noted that diasporas are not necessarily rooted in a desire to return, as a lateral and decentered attachment to a homeland may be equally as important and sufficient for the perpetuation of a particular collective (Clifford 1997).

Moreover, Brubaker (2005) points to "boundary maintenance" as a constitutive feature of diaspora populations (see also Armstrong 1976). Regardless of whether communities are preserved by deliberate resistance to assimilation in the host countries, or intended or unintended social exclusion from the larger societies their distinctiveness and group solidarity is maintained and mobilized across generations in host countries through the use of social boundaries. Rather than a specific physical place, diasporic populations draw on a unified national imaginary of social location to establish these ontological divisions, which typically involve a particular form of mobilization around cultural and/or religious heritage derived from the ancestral homeland (also Anthias 1998; Safran 2005, 2008). The existing sociopolitical, cultural, and economic connections to the historic land are also reflected in the symbols used as resources to support diasporas' imaginations, as well as in the communal social, cultural, and religious institutions established and maintained by communities in their hostlands. Stuart Hall builds on this point in asserting that members of a diaspora community not only share a history and sense of ancestry but also hold in common "one, shared culture, a sort of collective 'one true self,' hiding inside the many other, more superficial or artificially imposed 'selves'" (2003, 234). In this way, boundary maintenance allows diasporas to "recreate a culture in diverse locations" (Clifford 1994, 306), in addition to preserving their unique identity vis-à-vis a larger host society (or even societies) (Brubaker 2005).

Still, the globalized nature of diasporic populations means that their social boundaries and identities are often, and simultaneously, challenged by the host culture within which they are embedded. The diaspora experience is therefore characterized not by purity or homogeneity but, given the boundary-erosion that comes with transnationalism, is necessarily defined by difference, heterogeneity, and hybridity (Hall 2003). Clifford points to the inherent and oxymoronic tensions that exist between the preservation and destruction of diasporas, or the problem of the "changing same," wherein diasporic communities and their identities are "endlessly hybridized and in process but persistently there" (1994, 320). For this reason, diasporas from the same homeland but located in different hostlands could be, and often are, noticeably dissimilar due to the unique local dynamics in each country. Like this, a diaspora's shared sense of collective identity is not stable but, instead, a process which is never complete or a matter of "becoming" as much as of "being" (Hall 2003). As with the other identities that people uphold, membership within a diasporic community also subjectively influences a person's actions and worldview, including the ways they perceive and respond to other people, and, inherently, also the ways other people perceive and respond to them. The anecdotes found in the following pages of this chapter display this reality in showing how my hybridized identity as a member of the Ukrainian diaspora in Canada impacted my doctoral studies in Ukraine. As the chapter outlines, this was observed not only in terms of how people approached me while I conducted my research but also in how I came to understand myself as both a researcher and member of the Ukrainian diaspora.

Diasporic Identity as an Insider Attribute

The role of my diasporic identity in my research can be seen without an explicit discussion of my fieldwork: my interest in Ukrainian politics and nationalism stems from my own Ukrainian heritage and culture. Having grown up in a very active diaspora community in Canada, my formative years (up to the point when I began my doctoral studies) were spent greatly involved in Ukrainian ethno-cultural and religious activities (Safran 2005). Even whilst most members of the diaspora were born outside of Ukraine – as our family members migrated to Canada by choice or were exiled in the nineteenth and twentieth centuries – a "distinct community" (Clifford 1997) has continued across time and space in Canada through the preservation of traditional, cultural, and religious practices; this is demonstrated by the many Ukrainian institutions found in cities across the country such as dance groups, choirs, museums, churches, and cultural associations. The community's collective attachment to, and support for, Ukraine's independence and prosperity (Safran 2005) can also be seen through alliances both associated with, and separated from, the Canadian government, such as the Ukrainian Canadian Congress, and widespread support for the recruitment and deployment of Canadians as international election observers during Ukraine's presidential, parliamentary, and local elections. Grassroots activism, and especially initiatives in response to Ukraine's "sufferings" (Safran 2005) in recent years since the Euromaidan of 2013–2014, the beginning of the war in Donbas in 2014, and more recently with Russia's invasion on February 24, 2022 further exemplify the diaspora's sustained connection with their homeland. This is additionally evidenced by Canadian-led humanitarian efforts and the many proxy protests, demonstrations, and memorial ceremonies that have taken place in Canada, some annually, in response to (and even mirroring) sociopolitical events and commemorations happening in Ukraine.

While other migrant groups in Canada have similarly maintained a strong connection to their home states, the Ukrainian diaspora has nonetheless become a separate society (Armstrong 1976) as one of the country's largest ethnic groups constituted of approximately 1.4 million Canadians (Ukrainian Canadian Congress 2020). Although I cannot speak for other diaspora populations, nor for Ukrainian diasporas located in other countries, the community in Canada has very much maintained a collective memory and strong attachment to Ukraine across political boundaries based on certain cultural and religious narratives and myths about their homeland (Safran 1991, 2005). Canadian society's emphasis on multiculturalism has appropriately also allowed for the preservation and continuation of the diaspora's unique identity so that being across the world and, in some cases, never visiting Ukraine or even speaking the Ukrainian language, has not deterred a majority of Ukrainian-Canadians from regarding themselves as "Ukrainians." The level of involvement of members in the community indeed varies, as is true for all diaspora populations; however, even those minimally involved still typically pronounce their Ukrainian identity in certain circumstances, such as around the holiday season or when attending Ukrainian events. In this way, the idea of nationhood takes on an almost metaphysical significance for many members of the diaspora, despite differences in their strength of attachment to Ukraine (Reid 2015).

While I have always identified as Ukrainian-Canadian and am very familiar with Ukraine, having spent a significant amount of time there studying and volunteering prior to commencing my postgraduate studies, the four months I spent in the field conducting my doctoral research pushed me to greatly deliberate my identity and previously held beliefs about my culture and ethnicity. It must be noted that before I entered the field to conduct my doctoral research, I was very aware that my subjectivities would influence the data collection and overall progression of my project, especially as my association with the Ukrainian diaspora had previously positioned me as almost an "insider" within the communities I study. Still, I did not realize the full extent to which my ancestry would shape my interactions with the people in my field sites until I began data collection. In many cases, and as was demonstrated in the vignette at the beginning of this chapter, my ancestral ties significantly widened the scope of my project by introducing me to new people and information that had not previously been available. As was further exemplified in the above interaction, both the stateswoman and I had originally accepted my position as a researcher and hers as an informant – which was lucidly exhibited in the formality of, and lack of overt expressions of emotion during, the interview – yet, my admission of being a member of the Ukrainian diaspora almost immediately changed the dynamic and her overall approach to me. Specifically, this woman appeared to reassess the engagement upon learning this new information; the conversation was not merely between a researcher and a regional administrator but also between two women who shared Ukrainian heritage. Although I did not expect the woman to have prior knowledge about my diasporic ties, especially given that both my first and last names are English and we had corresponded in English via email to arrange the interview, seeing how drastically she changed her demeanor once she learned this fact revealed the significance of my subjectivities in this exchange.[2] Notably, I was not able to meet with the woman again when I returned to the same city six months later, however, she still arranged and recruited participants for four focus groups on my behalf in various rural locales across the region prior to my arrival. Our shared identity hence elicited a particular response from this regional administrator and, while it cannot definitively be determined, it is unlikely she has responded, or would respond, in the same way to non-diaspora scholars.

In other situations, my association with the Ukrainian diaspora in Canada garnered a sense of respect from my participants that allowed me – as a member of this community – to enter new sites within the field, including some of the more intimate places of their everyday lives. Exemplifying this was my experience in one village in a different region that I visited to conduct a focus group with retirees. Following the completion of the group discussion, my participants gave me a tour of the local museum and hosted a garden party with homemade food and drinks as a way for me to experience what they called "true Ukrainian hospitality."[3] Through these activities, I was able to naturally engage with local people and therefore learn more about Ukraine's culture, history, and politics from a grassroots perspective – the hosts also sent me home with a bag of fresh fruits and vegetables from their garden. As was likewise exhibited in the anecdote presented at the beginning of the chapter, the community members' perceptions of, and behaviors in response to, my Ukrainian heritage intrinsically shaped my research in this village by informing who and what I could access. Whilst I cannot assume that other scholars would not also receive the same cordiality from this community (or any of the others I visited), the excitement expressed by the local people about the fact that I had "returned home to Ukraine" clearly demonstrates that my ancestral ties motivated their actions, at least in some innate way.

Notably, I experienced similar responses and sentiments from my participants in different locales and regions across the country. For instance, I arrived in another small town expecting to conduct a focus group but was, instead, welcomed by a private tour of the town's gallery by the director, which was followed by an interactive session of several short presentations delivered by the region's most qualified experts on a variety of ethno-cultural topics. Once the focus group was conducted three hours later, I was abruptly ushered into a car and driven to another village, where I found more than 30 people waiting for me – the guest of honor – to begin their afternoon's festivities. During the evening of music, dancing, and food, I was introduced to various members of the community and even interviewed for a local newspaper about my impressions of their celebration and how it compared to the equivalent festival in Canada. From the perspective of a researcher, this event, like the many other unexpected dinner, concert, and party invitations extended to me, proved incredibly insightful for my research as an opportunity to interact with and observe my field sites and participants in an immersive and natural way. In these unstructured and unplanned situations, I was no longer trying to "control for biases" as I had been in the more formal exchanges earlier in the day (Fujii 2015), but was able to engage in the environment as both a participant observer and member of the Ukrainian diaspora. My experiences hence resembled those of Laura Adams' (1999) during her fieldwork in Uzbekistan, where her participants frequently required her attendance as a guest at various social occasions in order to show her "real Uzbek culture" (1999, 332). Whereas the invitations extended to me during my fieldwork were occasionally linked to the themes I study, just as Adams (1999) experienced, my diasporic identity nevertheless often proved more important in my interactions in the field than these particular topics.

Relatedly, my connection to the Ukrainian diaspora also opened up further observations and research opportunities to explore my field sites in a very intimate and astute, albeit unexpected, way during the COVID-19 pandemic. Since I was unable to physically visit Ukraine after I left in March 2020 due to national lockdowns and border closures, social media, and especially Facebook, became the primary means through which I was able to study Ukraine. Whilst I only casually mentioned in my initial messages to potential online participants that I was Ukrainian-Canadian by way of introduction (in addition to explaining that I was a doctoral student based in the UK studying Ukrainian politics), this particular detail seemed to be what stood out and motivated many of them to meet with me. In fact, several individuals who agreed to an interview used our conversation as an opportunity to ask their own questions about the Ukrainian diaspora in Canada. Demonstrating this is the following quote by a historian whom I interviewed online in the spring of 2020: "[y]ou were the first person from Canada who I talk to…the first person from Canada, and especially, from…Canadian-Ukrainian family."[4] Another participant asked if I would attend an online English-language club that she was involved with to speak about the Ukrainian diaspora. Others invited me to visit their homes once the pandemic was over, stating they wanted to show me the "real side of Ukraine" (this again resembles the experience of Adams 1999).[5] Although it cannot conclusively be discerned whether these comments were solely in response to my diasporic identity, my interlocutors nonetheless overtly exhibited much excitement about the fact that I identify as Ukrainian-Canadian (for more on this, see Howlett 2021). I have furthermore profoundly observed the impact of my diasporic ties since Russia's invasion into Ukraine in February 2022, as many friends and colleagues in Ukraine have reached out for assistance knowing I have wide networks within the diasporic communities in North America. When taken together, these stories thence portray the role that my diasporic identity played in shaping my interactions with others during my fieldwork (whether in-person or online) which, in turn, integrally influenced my overall data collection and the data themselves.

Importantly, I am aware that scholars, and especially junior female academics, are not advised to visit participants in their homes during fieldwork due to concerns around safety and security. Entering the field for the first time as a junior scholar can also prompt feelings of anxiety, and like all situations we find ourselves in – both in our home countries and in the field – ensuring our safety is of utmost importance. Before traveling to the above communities, like all others I visited, I therefore regularly considered the security risks and debated whether each situation was safe. Given my familiarity with the country, language, and culture, I never felt unsafe during my field research, but I recognize that this is not true for all scholars who conduct immersive fieldwork as there is indeed risk involved, especially for certain subsects of the population. Still, there were some instances during my doctoral research wherein I felt uneasy or uncomfortable engaging with participants in the field, particularly when I dismissed impromptu invitations and the generosity offered by my interlocutors, as it sometimes felt socially, and almost culturally or morally, disrespectful for me to do so. While not necessarily dangerous, these situations were sometimes socially awkward as I was required to navigate a fine line between ensuring my safety while also respecting my participants and their social norms.[6] My experiences during my four months in Ukraine thus echo several of the situations described in the literature on the methodological considerations involved in conducting fieldwork, particularly around the risks and ethics of immersive research, as well as highlighting firsthand the challenges not often discussed in prior works around balancing one's safety, research considerations, and socially appropriate behaviors and expectations while in the field.

Diasporic Identity as an Outsider Attribute

Although my diasporic identity allowed me to gain access to networks and start conversations that would have otherwise not been available to me – at least not in the same way – it also often created unforeseen methodological hurdles. While I had extensive fieldwork training and had consulted a great deal of methodological literature prior to traveling to Ukraine, my experiences still surprised me, especially as my ontological understandings of myself were at times challenged during my time in the field. In particular, I came to the profound realization that the ways I had previously understood myself as "Ukrainian" was actually a hybridized, diasporic identity. This reality was exposed by the fact that I rarely, if ever, felt a shared sense of identity with the communities or people whom I visited in Ukraine. Just as Safran explains with diaspora populations in their host country, I never truly experienced full acceptance by Ukrainian society during the time I spent in the country; instead, I always felt "partly alienated and insulated from it" (Safran 2005, 37). This sense of detachment came partially from the fact that I was visiting many new sites for the first time and because my direct interactions and exchanges with the people I met in the field lucidly underscored my position as an outsider. For instance, a man who became one of my closest friends during and following my time in Ukraine very explicitly reminded me, on several occasions, of my inability to socially integrate into Ukrainian society. Although not meant in a disrespectful or offensive way, he often told me that I did "not belong in Ukraine" as I was "too Anglo-Saxon." He also explained that although I dress in a way that is only slightly different from the local people, it is still noticeable enough for me to stand out – my backpack was one specific thing he said revealed that I was a foreigner. Several other participants also admitted they could visibly tell that I was "not from Ukraine," as I was too expressive and smiled too often to be a Ukrainian or Eastern European.

The fact that I was an "outsider" was further underscored through many conversations with an elderly individual I befriended in one of my research sites. Whilst our dissimilarities were evident from our first encounter, during which I fielded many questions about North America and my educational trajectory, in learning more about me and my background through follow-up meetings, the man began emphasizing our similarities as Ukrainians. However, as the dynamics between me (the researcher) and him (the participant) appeared to transform into a much more egalitarian relationship, a power imbalance was revealed in other ways. In particular, the man's questions about Canada and remarks about how "lucky" I was to have been born in North America accentuated that we did not occupy the same position. Although he simultaneously stressed that we are similar because he is Ukrainian and I am a member of the diaspora, in positioning me as someone "luckier" than him, the man inadvertently pointed to an unequal power dynamic in our relationship.[7] Although actively getting to know this man and being "brought into his world" (Tillmann-Healy 2003) allowed me to better understand the lives of my participants, and also exposed me to underlying sociopolitical dynamics in Ukraine, I was hyperaware of his perception of me during our exchanges as I oscillated between my identities as both a researcher and member of the Ukrainian diaspora (Finlay 2002). Being overtly reminded of my metacharacteristics, like I was by this man, thus fundamentally influenced my exchanges with participants in the field, whether consciously or not.

Like the story detailed at the start of the chapter about the regional administrator, interactions like this one again encouraged me to reflect on the significant influence of my subjectivities in the research process in order to find a balance between sharing many commonalities with my participants and objectively researching their country.

Evidently, then, while I was easily able to establish trust and arrange meetings because of my "insider" identity as a Ukrainian-Canadian, I was still intrinsically approached as an outsider by many participants. This ontological divide was particularly stark in situations when I spoke Ukrainian, as it was often disclosed that the vernacular I speak is "Canadian Ukrainian" and different from the language spoken in modern-day Ukraine.[8] For instance, as one politician explained to me during an interview:

> The Ukrainian that the Ukrainian diaspora speaks in Canada – this is the language they spoke at the beginning of the twentieth century. You have kept it. I feel you speak the language of Canadian Ukrainians. It is okay, you have kept it. But now, one hundred years have passed and Ukrainian language has changed. That's why Ukrainian you speak and Ukrainian I speak can be different, but still, it is Ukrainian.

These words were not exceptionally surprising to me given that I primarily studied the language in Canada and my ancestors, like many Ukrainian-Canadians, emigrated in the twentieth century for agricultural opportunities and in response to the political instability in Eastern Europe. Nonetheless, I was astonished by the frequency at which I was told similar things by the people whom I met across the country during my field research. In most of these instances, local people expressed how impressed they were that I spoke the Ukrainian language but stated they could easily identify me as a member of the diaspora based on my accent and vernacular. My use of Ukrainian therefore reinforced my diasporic identity and "outsider" characteristics in some situations, which sometimes proved uncomfortable as it highlighted a divide between my participants and I. At the same time, though, my experiences still showed the importance of knowing local languages, especially for native English scholars like myself working in non-English speaking environments, as a way to better connect with participants. As I observed during my time in Ukraine, knowledge of the local language(s) is extremely appreciated and shows respect for the communities and people we study. Since words and phrases do not always translate into other languages, having local knowledge and language skills of the regions we study thus allows for a more nuanced understanding of our participants' words, sentiments, and social realities.

In many of the encounters where our versions of the Ukrainian language were noticeably different, I experienced a shift in my understanding of the diaspora's culture and identity (and, inherently, also myself). Specifically, it was through these exchanges that I came to realize how the diaspora has reproduced and perpetuated a historical version of the Ukrainian language and culture in Canada, whilst the Ukrainian language and traditions in Ukraine have transformed and evolved in parallel, as is true for all diaspora populations.[9] In this way, the homeland that the Ukrainian diaspora in Canada (and in other countries) feels connected to is not actually Ukraine as it is in the modern-day, but an imagined construct based on historical narratives about the country as it was at a certain point in history, particularly when our ancestors emigrated. What is understood as "Ukrainian" in Canada, including certain customs and traditions, may therefore be (and likely is) different to how "Ukrainian" is understood in other hostlands, and different still from what is understood as "Ukrainian" in Ukraine.[10] Despite the cultural narratives and social practices perpetuated within the Ukrainian-Canadian collective emphasizing being "Ukrainian" first and foremost, the two cultures, identities, and societies remain distinct. Whereas my identity as Ukrainian-Canadian positioned me as an outsider in some situations during my field research, then, it was actually through being seen as an outsider that helped me to realize the ways being Ukrainian in the diaspora is different than being Ukrainian in Ukraine.

As such, my diasporic identity very much assisted me during my field research in gaining access to some sites and establishing rapport with participants as an individual with inside knowledge about the sociocultural context of Ukraine, yet, it paradoxically also defined me as an outsider and someone noticeably different from those with whom I interacted. At the same time that being a member of the diaspora proved sufficient for me to be welcomed by and given access to Ukrainian society, it was still not enough for me to fully integrate into the local communities I study (Clifford 1994). Being positioned in this gray or 'fuzzy' way often left me with a confused sense of self in feeling like both an insider and an outsider in a society that I had previously thought I understood and even felt attached to at least prior to conducting my field research. In struggling to reconcile my bifurcated identity, I remarkably came to experience a similar sense of unease in the *homeland* that Safran (2005) explains diasporic populations typically feel within their *hostland.* In many ways, my experiences as a Ukrainian-Canadian conducting field research in Ukraine were, in fact, comparable to those of diaspora populations living in their hostlands; specifically, my lateral and decentered attachment to Ukraine helped me relate to the country and my participants on a very personal level, while my Canadian citizenship prevented full acceptance (Safran 2005). Although citizenship does not equate to diasporic identity, my simultaneous attachments to two political entities (Clifford 1997) led to an existential predicament as I navigated these complex dynamics throughout my fieldwork.[11]

Moreover, as I had felt comfortable with Ukraine and its culture prior to commencing my doctoral studies, it sometimes proved challenging for me to separate my curiosity as a member of the diaspora from my academic pursuits. Even though I was aware of the importance of reflexivity and the need to acknowledge our subjectivities – particularly because of the extensive methodology training and literature I had consulted prior to fieldwork – I was still somehow, and somewhat problematically, unaware of how significantly my ancestral heritage could risk complicating my research. I had neither expected my fieldwork to be particularly taxing, at least not in the emotional way that it was, nor that I would leave the field integrally questioning my own identity. In this way, the personal and ontological trials that I faced, in addition to the interactional dynamics associated with my diasporic ties, not only forced me to deeply reflect on my data collection processes and doctoral research project more largely but also on the complexities and even difficulties sometimes involved in conducting transparent, ethical, and sound qualitative research. My field research was therefore a multifaceted learning experience that not only taught me about the people and places I study but, fundamentally, also myself.

Concluding Thoughts

Peregrine Schwartz-Shea writes that being reflexive involves "a keen awareness of, and theorizing about, the role of the self in all phases of the research process" (2013, 133). Beyond self-assessment and reflection, reflexivity encourages researchers to gaze both outwards and inwards, and consider the taken-for-granted privilege and power structures embedded within the relationships we establish in the field (Fujii 2015). Throughout my own ethnographic research both physically in Ukraine and online during the COVID-19 pandemic, the act of being reflexive helped me become more aware of the ways I made sense of unstructured and unexpected observations and happenings, such as those outlined in this chapter. Although the interactions with my participants contributed to the ontological anxieties that I felt in some situations, reflexivity helped me to continuously and critically examine my intervention in Ukrainian society as a member of the diaspora (Probst 2015), including all of the associated and resulting meanings and power dynamics (Fujii 2015). These "internal dialogues" (Berger 2015) consequently aided me in becoming much more aware of my own sense-making and ensured greater accountability, rigor, and integrity in the progression of my doctoral research. In addition, actively reflecting on my biases and subjectivities allowed me to better interact with my field sites and research participants through a more nuanced investigation in recognizing idiosyncrasies and subtleties I may have otherwise missed. While direct engagement with my participants was fundamentally curtailed during the pandemic, my online research still demonstrated the influence of my subjectivities, specifically my identity as a Ukrainian-Canadian, in allowing me to observe and better understand particular aspects of my participants' worlds on the ground (Howlett 2021).

Importantly, and as the above reflections demonstrate, my diasporic identity is not the only ascribed trait which motivated my participants' actions and behaviors. Being a young female in particular raised ethical and methodological considerations at certain times throughout my fieldwork, especially given the sociocultural understandings of gender within former Soviet states. In addition to many queries about my marital status, and comments about the fact that I was spending a significant amount of time traveling alone in Ukraine, I was often invited to private or intimate meetings with interlocutors, typically men. In one instance, for example, a male participant asked me to meet and practice his English over dinner on Valentine's Day.[12] Although I politely declined his invitation, it remains unclear to me whether male scholars would be put in similar positions or even granted the same access within the field if they were to carry out a similar research project. This situation, like some of the others described in the chapter, points to the potential risks associated with being a solo female scholar conducting overseas field research, and shows how quandaries may arise – and even become ethically dubious – when societal expectations clash with the bureaucratic procedures intended to ensure ethical research. Whereas the unstructured and exploratory nature of ethnographic fieldwork can indeed produce rich data and in-depth observations, as was very much the case in my own research, the fact that we cannot foresee or plan most of the situations wherein we may find ourselves in the field is hence a hurdle implicit to the method of fieldwork that must be navigated by scholars of all genders and levels of experience.

Yet, as the anecdotes presented in this chapter illustrate: the ways one's positionality and ascribed identities implicate their exchanges with others and, subsequently also others' perceptions and responses to them, are both an unpredictable and inevitable part of fieldwork. This reality is also not isolated to research settings, but true in any social context. Whilst we may not always realize the significance of our positionalities, it was evident to me when I conducted my doctoral research. Even though my ancestral connection to Ukraine more often aided my research, it must be noted that our subjectivities equally create limitations for research. Namely, my participants' actions affected the reproducibility of my findings as it is unlikely that other scholars will be able to carry out similar studies (Schwartz-Shea and Yanow 2011). Moreover, it is very improbable that non-diaspora scholars would be able to observe my participants' everyday lives in the way that I was fortunate to both in-person and online, especially going forward in a post-pandemic world and in post-war Ukraine. While it can indeed be unexpectedly challenging, stimulating, and unstraightforward to navigate our various traits and subjectivities in our research, especially for junior scholars, my experiences in Ukraine accordingly highlight that self-conscious reflection can ensure more ethical and transparent data. Furthermore, and as I hope this chapter has shown, it may also be through active reflection that we come to better understand ourselves as both researchers and social beings.

Notes

1 I would like to acknowledge the many individuals who both directly and indirectly contributed to the writing of this chapter. In particular, a great deal of credit is due to my participants – those whose stories appear here and equally those whose do not – as you have collectively helped me to understand Ukraine, my field sites, and also myself in a new way. Additional thanks to Marissa Kemp and Rachel Valbrun for helping me think through the intricate ideas explored here, and to Viktoria Sereda, Elizaveta Potapova, and Katria Tomko for their insightful comments on earlier versions of this chapter.

2 It must be noted that not every participant was aware of my diasporic ties; therefore, it cannot be assumed that their responses and behaviors were always because of this particular aspect of my identity. Nonetheless, the exchange with this stateswoman made me much more aware of how all participants could perceive me because of the positions that I occupy.

3 Focus group conducted on July 14, 2019.

4 Interview conducted on June 3, 2020.

5 With this reference, my participants covertly indicated that Ukrainian culture in the diaspora is not "real" Ukrainian culture.

6 As food and the sharing of meals is central to Ukrainian culture, turning down such hospitality – a social taboo – could be interpreted as rude and potentially risk my reputation with participants, including even limiting future access to that particular field site.

7 While it was never explicitly stated what was "luckier" about my life in North America, the themes discussed in the conversation very much pointed to the fact that the man associated North America and higher education with great financial and intellectual wealth.

8 Notably, the Ukrainian language spoken in Canada has (and still does) vary based on the different waves of immigration and where migrants emigrated from. For example, Western Ukraine's linguistic patterns have been significantly impacted by the Austro-Hungarian Empire and Poland, whereas Eastern Ukraine's are more influenced by Russian/Soviet rule. The current Ukrainian-Canadian dialect arguably reflects the World War II wave of immigration rather than those who emigrated more than a century ago. This is because most Ukrainian-Canadians associated with the first wave no longer speak Ukrainian at home (at least as the dominant language), my family included, while those who do have mostly learned it at Ukrainian schools which have been revived and heavily influenced by the World War II wave of immigration.

9 One popular vernacular spoken throughout Ukraine, and in many of my field sites, is a mix of Russian and Ukrainian called *Surzhyk*.

10 Interestingly, I came to realize through my interactions with participants that Ukrainian society's understanding of the diaspora, especially the community in Canada, is equally based on historical narratives and myths (Safran 1991, 2005). In this way, both Ukrainian society and the Ukrainian diaspora have similarly constructed the other – and based their perceptions on these understandings – independently and without significant knowledge about the realities of the corresponding populations.

11 It must be noted that not all situations and interactions with participants in the field challenged my self-understandings.

12 Although Valentine's Day is not a significant holiday in Ukraine, I know this man was very aware of the holiday and its underlying meanings as he and I had attended an English-speaking club earlier in the week where the meeting's theme was "Love and Valentine's Day."

References

Adams, L. 1999. "The Mascot Researcher. Identity, Power, and Knowledge in Fieldwork." *Journal of Contemporary Ethnography* 28 (4): 331–363. DOI: 10.1177/089124199129023479.

Anthias, F. 1998. "Evaluating Diaspora: Beyond Ethnicity?" *Sociology* 32 (3): 557–580. DOI: 10.1177/0038038598032003009.

Anthias, F. 2008. "Thinking through the Lens of Translocational Positionality: An Intersectionality Frame for Understanding Identity and Belonging." *Translocations: Migration and Social Change* 4 (1): 5–20.

Armstrong, A. 1976. "Mobilized and Proletarian Diasporas." *American Political Science Review* 70 (2): 393–408. DOI: 10.2307/1959646.

Berger, R. 2015. "Now I See It, Now I Don't: Researcher's Position and Reflexivity in Qualitative Research." *Qualitative Research* 15 (2): 219–234. DOI: 10.1177/1468794112468475.

Brubaker, R. 2005. "The 'Diaspora' Diaspora." *Ethnic and Racial Studies* 28 (1): 1–19. DOI: 10.1080/0141987042000289997.

Cambridge Dictionary. 2020. Cambridge: Cambridge University Press.

Clifford, J. 1994. "Diasporas." *Cultural Anthropology* 9 (3): 302–338. DOI: 10.1525/can.1994.9.3.02a00040.

Clifford, J. 1997. *Routes: Travel and Translation in the Late Twentieth Century.* Cambridge: Harvard University Press.

Finlay, L. 2002. "Negotiating the Swamp: The Opportunity and Challenge of Reflexivity in Research Practice." *Qualitative Research* 2 (2): 209–230. DOI: 10.1177/146879410200200205.

Fujii, L. 2015. "Five Stories of Accidental Ethnography: Turning Unplanned Moments in the Field into Data." *Qualitative Research* 15 (4): 525–539. DOI: 10.1177/1468794114548945.

Fujii, L. 2016. "Politics of the 'Field.'" *Perspectives on Politics* 14 (4): 1147–1152. DOI: 10.1017/S1537592716003236.

Hall, S. 2003. "Cultural Identity and Diaspora." In *Theorizing Diaspora: A Reader*, edited by J. Braziel and A. Mannur, 233–247. Maiden: Blackwell Publishing.

Howlett, M. 2021. "Looking at the 'Field' through a Zoom Lens: Methodological Reflections on Conducting Online Research during a Global Pandemic." *Qualitative Research* (January): 1–16. DOI: 10.1177/1468794120985691.

Probst, B. 2015. "The Eye Regards Itself: Benefits and Challenges of Reflexivity in Qualitative Social Work Research." *Social Work Research* 39 (1): 37–48. DOI: 10.1093/swr/svu028.

Reid, A. 2015. *Borderland: A Journey through the History of Ukraine.* London: Weidenfeld and Nicolson.

Safran, W. 1991. "Diasporas in Modern Societies: Myths of Homeland and Return." *Diaspora: A Journal of Transnational Studies* 1 (1): 83–99. DOI: 10.1353/dsp.1991.0004.

Safran, W. 2005. "The Jewish Diaspora in a Comparative and Theoretical Perspective." *Israel Studies* 10 (1): 36–60. https://www.muse.jhu.edu/article/180371.

Schatz, E. 2009. *Political Ethnography: What Immersion Contributes to the Study of Power.* Chicago: University of Chicago Press. DOI: 10.7208/9780226736785.

Schwartz-Shea, P. 2013. "Judging Quality: Evaluative Criteria and Epistemic Communities." In *Interpretation and Method: Empirical Research Methods and the Interpretive Turn*. 2nd edition. edited by D. Yanow and P. Schwartz-Shea, 89–114. Armonk: M.E. Sharpe.

Schwartz-Shea, P. and D. Yanow. 2011. *Interpretive Research Design: Concepts and Processes*. New York: Routledge. DOI: 10.4324/9780203854907.

Tillmann-Healy, L. 2003. "Friendship as Method." *Qualitative Inquiry* 9 (5): 729–749. DOI: 10.1177/1077800403254894.

Tölölyan, K. 1991. "The Nation-State and Its Other: In Lieu of a Preface." *Diaspora* 1 (1): 3–7. DOI: 10.3138/diaspora.1.1.3.

Ukrainian Canadian Congress. 2020. "Ukrainian Canadian Congress." Available at: https://www.ucc.ca (accessed on October 20, 2020).

Wood, E. 2009. "Field Research." In *The Oxford Handbook of Comparative Politics*, edited by C. Boix and S. Stokes, 123–146. Oxford: Oxford University Press. DOI: 10.1093/oxfordhb/9780199566020.003.0005.

Yanow, D. and P. Schwartz-Shea. 2014. *Interpretation and Method: Empirical Research Methods and the Interpretive Turn*. 2nd edition. Armonk: M.E. Sharpe.

Part III

Stories from the Digital Field

6 Listening and Its Limits

Reflections on Fieldwork in/on Kyrgyzstan

Colleen Wood
Columbia University

Introduction

The ethnographer's body is her primary instrument for research: her senses translate the field to help her brain recognize what she sees, hears, and tastes; her social body shapes her relationship to the field and filters the automatic process of sensing, synthesizing, and understanding. Some scholars frame listening and hearing as two ends of a spectrum, distinguished by the degree of intentionality when taking in information with listening as a more *active* verb, but this spectrum masks a much more diverse range of listening practices that researchers can hone. We can listen intently, deeply, humbly (Koch 2020); we can eavesdrop, strain to hear, or take in conversations while listening to music, watching the news, and driving.

In this chapter, I reflect on three periods of fieldwork between 2015 and 2022 and explore the range of listening practices that I used while conducting research in Kyrgyzstan. I specifically consider how my positionality, as a young female scholar from the United States, shaped my field encounters and epistemological commitments. I begin by recounting my experiences as a Peace Corps volunteer in Kyrgyzstan, where I was taught that vigilant listening was a matter of personal and national security. Next, I discuss the process of being disciplined within the field of political science, learning to make sense of multiple voices with disparate levels of academic and customary authority. Finally, I reflect on my experiences of learning to listen while conducting fieldwork on civil society in Central Asia from afar, mediated by screens and the strength of a Wi-Fi signal – a task that is still in progress as I complete my dissertation from the United States as a consequence of the COVID-19 pandemic's impact on global mobility.

By reflecting on these three periods of my field experience in Kyrgyzstan, I identify several parallels between fieldwork training and what Heathershaw and Megoran call a "discourse of danger" (2011, 589). According to these scholars, the West's geopolitical gaze "derive[s its] claims and contentions in accordance with a preconceived and self-referential discourse of danger that identifies threats to [W]esterners while failing to appreciate the insecurities that are felt and experienced by Central Asians" (Heathershaw

DOI: 10.4324/9781003144168-10

and Megoran 2011, 594). Hence, to Western policymakers and journalists, Central Asia's obscurity, orientalism, and fractiousness make it a "dangerous" place to live, work, and study. Importantly, Western academics can also fall into the trap of reproducing these narratives in fieldwork training, guides, and advice. In this chapter, I unpack and reinterpret these supposed traits of Central Asia as an area of study while calling on the academy to double down on efforts to listen to colleagues and interlocutors from the region.

Learning to Listen for Personal and National Security

I became interested in Central Asia as a high school student when I took my first formal Russian class with a Kyrgyzstani teacher. My curiosity soon grew into a fascination with Central Asia's history and politics, and I found ways to craft my final papers around the region during my undergraduate studies. After graduation, a two-week trip to Kyrgyzstan made a deep impression on me, and I landed again in Bishkek six months later with a cohort of 60 Peace Corps volunteers, ready and eager to spend more than two years volunteering in the country. My motivations for applying to the Peace Corps – a fixture of Cold War-era foreign policy, which has shipped thousands of Americans around the world since 1961 to serve rural communities in the name of cross-national collaboration and friendship – were hardly ideological. I was more motivated by the opportunity to learn Kyrgyz than anticipating any success as a development worker.

For the first seven weeks in Kyrgyzstan, my cohort spent several hours each day in Kyrgyz language classes in addition to training in pedagogy, how to conduct needs assessments, and maintaining personal security in the field. A discourse of danger – similar to the one Heathershaw and Megoran (2011) described at the heart of foreign policy and journalistic writing on Central Asia – was implicit in our training, which covered everything from healthy eating to professional success as volunteers. Advice given to Peace Corps volunteers on how best to maintain personal and national security specifically hinged on a discourse of Kyrgyzstan being "a fractious place" (Heathershaw and Megoran 2011, 590). In our training courses, the potential for ethnic conflict and widespread misogyny were the primary manifestations of this fractiousness.

After horrific interethnic violence in southern Kyrgyzstan in June 2010, the Peace Corps shut down operations in Osh and Jalalabad provinces for five years. I was one of the first volunteers to be sent to Jalalabad after the June events, and the (mostly ethnic Kyrgyz) staff were anxious about living in ethnic minority communities. On a check-in call a few days after settling into my host family's home, a Peace Corps staff member warned me to be careful about walking around alone because I lived in a Uzbek neighborhood. Her unsaid implication was that my proximity to ethnic minorities – especially in southern Kyrgyzstan – put me at risk, although it was never clear whether that risk stemmed from the possibility of renewed interethnic violence or prejudicial beliefs about minorities. As volunteers in southern Kyrgyzstan we were coached to be particularly vigilant about our surroundings and, unlike those living in northern regions, we were temporarily evacuated from our field sites during elections and other politically tense moments.

In addition to removing volunteers in southern Kyrgyzstan from their sites, the Peace Corps pushed immersion and language skills as strategies for maintaining personal and national security. Concrete advice on how to stay safe was often emphasized for individuals with membership in certain marginalized groups, especially women. Much of our training centered on how women could avoid being assaulted and raped, with role-playing activities to practice avoiding dangerous situations at night clubs, at work, on public transportation, and in our host families' homes.

The Peace Corps staff wanted its volunteers – especially women – to be astute listeners with strong language skills. Volunteers in my cohort therefore received almost 200 hours of structured language training in Kyrgyz before being sworn into the organization. Of course, language study is an end in and of itself, but having a strong grasp of the language served security purposes to help people avoid and get out of challenging or potentially dangerous situations. In one session, we were split by gender in one language class for a session on curse words, during which our normally staid teachers wrote words like "slut" and "dick" on the white board. They scolded us for laughing, reminding us of the need to be vigilant, as hearing words about sex or marriage could signal danger, and we needed to remove ourselves from situations where people talked like this. But this advice proved difficult to follow in the field. Once, in a five-hour taxi ride to the capital, the driver asked me whether I had heard of the "tradition" of *ala-kachuu* (bride kidnapping, literally "take and run" in Kyrgyz). His tone was joking but not without a soft undertone of the possibility of sexual violence. As a foreign woman, I knew I would not actually be forced into marriage in the way so many girls and young women in Kyrgyzstan are, despite the practice having been outlawed for many years (Margolis 2015). Even so, how was I supposed to remove myself from the situation, where was I supposed to go? My female coworkers traded quips to respond to suggestive jokes: some tried humor (that they were already married or that they would be a bad wife because they could not milk cows or make bread), some tried polite albeit more blatant approaches (curtly stating that Kyrgyzstan's Criminal Code prohibits bride kidnapping and is punishable with five years in prison), and some just stayed silent. Unsurprisingly, these tactics landed differently depending on the day, the mood of the joke-giver, and the mood of the volunteer, suggesting the futility of supposedly foolproof advice for navigating the field safely.

The emphasis on the danger of sexual assault was well intentioned, given persistently high rates of gender-based violence for Kyrgyzstani women (Eshaliyeva 2020) and because Peace Corps volunteers serving in

Kyrgyzstan have historically experienced sexual harassment and rape more frequently than in other posts worldwide (Peace Corps Office of Inspector General 2019). While an awareness of one's surroundings, heightened by gender and phenotypic markers, is certainly important for researchers when in the field, I contend that stressing the constant danger of gender-based violence actually served to exoticize the country, making an image of Kyrgyzstani men as particularly predatory or impulsive. This framing of universal dangers as though they were unique to Kyrgyzstan furthermore reinforced our American assumptions and heightened our fears, often making volunteers more fearful than they needed to be. Moreover, the emphasis on the potential dangers that women may face in the field fundamentally ignored the reality that catcalling, assault, and rape also exist – and are in fact very common – in the United States.

Although aspects of our training perpetuated Heathershaw and Megoran's "discourse of danger," the emphasis on language acquisition and immersion taught me invaluable listening skills that later served me as a social scientist.

In the early months of my time in Kyrgyzstan, before I had consolidated my grasp of Kyrgyz, listening was my default approach to socializing with others. When attempting to speak with my neighbors or host family, I would concentrate on preparing grammatically correct sentences in my head by rearranging the agglutinative language emblematic of Turkic languages until I thought I had it correct. By the time I was ready to announce my contribution to the conversation, though, the discussion had moved on. Consequently, I realized that it made more sense to just *listen* to the people around me and jot down new words from context in a tiny black notebook that I carried with me all over the country. Unlike most Peace Corps volunteers, who spend their entire service in one place, I did not have a single "home base" during the two years that I lived in Kyrgyzstan; instead, families across the country generously opened their homes to me. First, I completed my initial training in a village about an hour's journey from Bishkek, Kyrgyzstan's capital. I then spent about a year in Jalalabad – a large city on Kyrgyzstan's southwest border with Uzbekistan – before moving to a resort town on the shore of Issyk Kul, a massive alpine lake known as "the pearl of Kyrgyzstan." I lived in Issyk Kul for a year before moving to Osh, Kyrgyzstan's second largest city, for the final four months of my service. As I moved around the country, I absorbed elements of local dialects, but my first year in the south infused my Kyrgyz with more Persian vocabulary than the capital city's vernacular. For instance, by dipping my bread in honey I called *asel* instead of *bal*, and by addressing taxi drivers as *ake* instead of *baike*.

Picking up these mannerisms was one form of immersion, which was presented as another key component of volunteers' personal security while in the field. The logic pushed by the Peace Corps is that if people know and recognize you, the less likely they are to want to hurt you. Whether or not this was accurate, I dove headfirst into Kyrgyz culture. I introduced myself to new people as "Kaliya," a Kyrgyz name my host mother gave me when we were crammed in the backseat of a Lada with five adults and two children on my first night in Kyrgyzstan. I memorized a long passage from the half million-line Manas epic to present for a holiday concert, mimicking my host aunt's intonation and hand gestures, learning to make my voice deep and vibrate with the first *Eeeiiiiii* that every performer uses to open their recitation of the poem. I tried to dress modestly, and, on special occasions, wore velvet vests decorated with ornate traditional stitching. But even though I embraced "Kyrgyz-ness" to the best of my abilities for myself, my host family, colleagues, strangers on public transport, and even for national TV audiences, I ultimately remained a foreigner. Sometimes, I passed for Russian, but it never took long for strangers to reassess me – usually a function of my muddy shoes and unironed dress that embarrassed my host mother, and my North American tendency to smile at everyone I pass on the street.

The surface-level immersion of changing my clothes made the experience of locating and pushing the boundaries of membership categories feel like exploring a new room in the dark. It took deeper listening to better appreciate the social, political, and economic contexts where I lived. But I could not ask about these contexts outright, as three topics are verboten for Peace Corps volunteers during their service: partisan politics, the military, and religion. I understood these boundaries on acceptable conversation to be protective mechanisms on multiple scales. First, they insulate volunteers from finding themselves in hostile conversations without the necessary vocabulary to navigate; second, they protect the local Peace Corps post from accusations of harboring spies; and third, they keep the United States' image safe from off-the-cuff riffing about our country's role in international politics. However, given that the topics most locals want to talk about over tea (or a bottle of vodka) are often those which are off-limits, it is extraordinarily difficult in practice to avoid these conversations.

As no rules prevented me from sitting with local people who happened to be discussing these issues, I spent a significant amount of time listening to conversations about political and social life. Eventually, I figured out how to probe my communities' thoughts on these issues without directly asking them or offering my own opinions. What strikes me now is not only the intentionality of my listening but also the vigilance with which I approached taking in noises and words around me. I listened to my host siblings' dinner table jokes and pop culture obsessions, my host parents describe their ancestors' histories and hopes for their children's futures, and my students worry about getting a meaningful education and finding work in a struggling economy. Two years of careful listening as a Peace Corps volunteer equipped me with interviewing skills and deep substantive knowledge that would later serve me well as a social scientist in training.

Learning to Listen as a Social Scientist

Immediately after completing my Peace Corps service, I began a PhD program in political science at Columbia University. The move from Osh – with a population of 250,000, a relative metropolis compared to some of the villages where I had lived – to New York City was jarring, and the transition from days with little structure to endless reading lists was exhausting and gratifying. Along with training in statistics and formal modeling, I sought opportunities to learn the ins-and-outs of qualitative and interpretive methodologies. Through my doctoral coursework, I gained a vocabulary to describe the way I had learned to approach my life and relationships in Kyrgyzstan during my years in the Peace Corps. Immersion was not only a goal for Peace Corps volunteers but also a tenet of ethnographic data collection and analysis.

My time in the field and earlier ethnographic experiences were assets in graduate school, but I often struggled to justify my geographic focus on Central Asia more than peers who study more "legible" regions (legible here meaning familiar in American geopolitical imaginings). While research on Russia, China, or the United States is defended by geopolitical dominance, research on Central Asia requires several paragraphs explaining how and why insights from these cases lend themselves to more general theories. This speaks to another component of Heathershaw and Megoran's "discourse of danger." They argue that Central Asia is often framed as obscure because it is "particularly distant, inaccessible, and unintelligible" (Heathershaw and Megoran 2011, 594). In addition to the pressure to justify my case selection and explain where Kyrgyzstan is to tenured faculty, this obscurity manifested as a gap between the neatly articulated hypotheses and concepts of canonical texts and my lived experiences in Kyrgyzstan.

While discussions of ethics often center around methods in political science, Koch (2020) reminds scholars that they should also question their complicity in reproducing problematic narratives through the selection of topics and the perpetuation of certain theories that presume to understand a place better than those actually living there (Koch 2020, 5). When I started graduate school, I was interested in scholarship on the link between language, identity, and education; I was struck by what I saw as a strict conceptual binary between civic and ethnic nationalisms in the literature.

Just as Koch questioned taken-for-granted academic concepts like "dissent" and "agency" in her research on autocratic regimes, I pushed myself to query dominant narratives within the study of nationalism and instability in Central Asia. My first fieldwork experience as a PhD student was in Kyrgyzstan, where I conducted interviews with teachers and mid-level bureaucrats in the Ministry of Education about civic education and language policy in schools.[1] The hours I spent speaking with interviewees and observing school hallways and Ministry of Education outposts revealed a much more complicated story than a simple shift from the "civic" nationalism that marked politics in the 1990s to potentially violent "ethnic" nationalism in the early 2000s (Eurasianet 2013). Although the number of Uzbek-language schools decreased following pogroms in southern Kyrgyzstan in 2010, I found through my qualitative research that material limitations and a sense of responsibility to care for pupils' civic consciousness pushed teachers to deviate from the state-sponsored curriculum. If I had not listened against the grain of the literature on nationalism and the Kyrgyzstani governments statistics, I would not have understood the complexity of the language-identity-education nexus as it is experienced and reproduced in southern Kyrgyzstan. No doubt, inequalities in education among ethnic groups exist in Kyrgyzstan and are worth careful analysis,[2] but the work of listening to informants describe their worlds pushed me to question the meta-narratives that dominate the study of nationalism in Kyrgyzstan and other post-Soviet countries, specifically the tendency toward zero-sum framing between civic and ethnic nationalisms.

Despite winning an award at a conference for a paper I wrote based on this fieldwork, I felt nervous about contributing to the scholarly debate on the dynamics of post-Soviet nationalism. While reading scholarship based on impressive, extensive field experience from the early 1990s and 2000s,[3] I struggled to understand what these cities, roads, borders, families, and schools looked like in the decades before I first visited. I was born following the collapse of the Soviet Union, graduated from high school in June 2010 – the same month as the spiral of interethnic violence in Osh – and first visited Kyrgyzstan several years after the violence ended. Given that I "missed out" on the formative years of Kyrgyzstani political and social development, I questioned as a young scholar what insight I could possibly offer on nationalism, state capacity, and violence within this context.

At that point, I had only thought about my age from the perspective of its intersection with my gender as these social identities shaped my access to the field and affected my relationships with those who assist in knowledge production processes, as interpretivist methodologists have theorized (Sirnate 2014; Fujii 2017). Many researchers have written about the challenges unique to being a young woman conducting research in patriarchal contexts (Nilan 2002; Johansson 2015; Kapiszewski, MacLean and Read 2015). Other scholars have considered how different social identities can complicate fieldwork, including race, ethnic presentation, sexuality, and disability (Ortbals and Rincker 2009; Behl 2017). But interrogating dominant academic narratives calls for reflexive listening in the field, not only in the sense of honestly and humbly considering what informants choose to share with us but also by reflecting on which words spark "aha" moments of conceptualization and theorization. I contend that age is more than a social identity that shapes the relationship between a researcher and her informants, and that youth can offer analytical advantages. Age functions as a bound on a scholar's possible store of "headnotes," what Emerson, Fretz, and Shaw conceptualize as the memories and thoughts that settle back in the deep corners of our mind until they are made relevant much later, perhaps on repeat visits to the field or while synthesizing and analyzing our field notes and jottings (2011, 23).

While the scholars whose work I admire visited Kyrgyzstan decades before I arrived, my age and the timing of my formative field experiences have informed my theoretical and empirical insights. For example, I have always known Kyrgyzstan through a smartphone, sending voice notes on WhatsApp because it is faster to speak due to the Russian keyboard's autocorrecting of Kyrgyz words, and resting easy in the comfort of digital connections in Kyrgyzstan and back in the United States when help may be needed. Reading academic texts published in the 2010s that advanced a skeptical view of the potential for telecommunications technology to affect political outcomes (Beissinger 2017), I was confused by the assertion that social media platforms were apolitical spaces and that few average people were connected to the Internet (Anceschi 2015). My headnotes and impressions of social and political life in Kyrgyzstan made my perspective on the power of social media different than the pessimism that has traditionally defined scholarship on this question, allowing me to make a substantial contribution to the ongoing debates (Dall'Agnola and Wood 2022).

Learning to Listen from Afar

In early 2020, prospects for international fieldwork shrank as the COVID-19 virus spread across the world and governments closed their borders. I was initially frustrated by the prospect of writing my doctoral dissertation on civic activism in Central Asia without being able to return to the region. With the limits on global travel for Americans only temporary, though, it would be unfair to say I fully understand the restricted mobility that has been the standard for many researchers without passport privilege. Recognizing the privilege of assumed mobility for scholarship, I decided to rework my project rather than wait indefinitely for borders to open.

The inspiration for reworking my dissertation came in October 2020 when I retweeted an image of crowds that had gathered in Bishkek's central Ala Too Square to protest parliamentary election results. By dawn, protesters had taken over the executive building in Bishkek, stopping to brew tea between destroying portraits of elected officials and setting fire to the fourth floor. In the days that followed, I hardly slept, waking up before dawn each morning to "go to Bishkek" via live stream. With my desktop, laptop, tablet, and phone all plugged in, I was able to stretch my eyes, ears, and mind across platforms, time zones, and languages. I translated YouTube videos, Instagram posts, and Facebook threads for those trying to keep up with developments from significant distances. I may not have been marching with crowds in Bishkek (indeed, if I had been in Kyrgyzstan, it would have been imprudent to join the rallies, as common criticism of these protests is that they are funded by Western forces that want to destabilize the country), but my experience of the events was still real and physically embodied. It was disorienting to close my laptop at the end of a day of "fieldwork" and realize that I was not actually in Bishkek but in my apartment in Manhattan. Having attuned my ears to code-switching between Kyrgyz and Russian chants on Instagram Live videos of the rallies, it was jarring to hear the Caribbean lilt of my neighbors' Spanish and nearby ambulance sirens that reminded me of the pandemic's persistence in New York City.

The work of watching and processing Kyrgyzstan's political crisis in October 2020 was a quick but thorough introduction to digital ethnography. In shifting my mindset from being "when" instead of "where" during my research (Gray 2016, 504), I realized the immersion and listening practices I had honed over several years could still serve me well, even from 10,000 miles away. Of course, the listening that grounds digital ethnography has a different dynamic than the practices that define face-to-face interviewing or observation. When speaking with someone in person, my social body – a function of my gender, age, nationality, ethnic presentation, and mannerisms – shapes the way others interact with me (Falconer al-Hindi and Kawabata 2002; Moss 2002; Ackerly and True 2008). I therefore understand reflexive listening to require thoughtful consideration of the significance of the words I hear from the people I engage with. Specifically, the speaker may choose certain words over others for many reasons: they think I want to hear them (social desirability bias); they assume I cannot understand (as often happens when I meet someone new and they ask my host in Kyrgyz who I am); they want to obscure the truth (though lies are also data, as Fujii 2010 and Allina-Pisano 2009 artfully point out); or they just want to get a rise out of or provoke me.

That is not to say that we lose our social body or our ability to listen reflexively, when conducting research online. There is a rich literature on the ways virtual interactions are still embodied and embedded in social structures (Sanjek and Tratner 2015), and evidence shows that women and minorities suffer more from trolls and digital harassment (Veletsianos et al. 2018; Are 2020). Yet, being in a position to "lurk" by reading threads without posting or engaging changes much about the way field interactions take place (Schrooten 2015; de Seta 2020, 84). In particular, lurking means that what people say and how they say it is no longer a function of *me* being in the audience as it is with in-person fieldwork. While the ways my interlocutors perceive my identity and research goals shape our direct digital interactions, this is not the case when I watch and sometimes screenshot Instagram stories of human rights organizations, activists, and journalists.

The modality of digital ethnography makes the volume of information to process substantially different compared to in-person ethnography. If we think of traditional ethnographic interviews as drinking a cup of tea, relying on digital ethnographic methods to follow political unrest and public opinion across a half-dozen social media platforms methods can sometimes feel like holding one's face to a fire hydrant. Although it can be overwhelming, this type of research offers several advantages. For instance, different apps' affordances offer researchers a range of ways to interact with data (Tufekci 2017). A researcher can virtually attend a protest as it is live streamed on Facebook and simultaneously take field notes about the stream of comments, the events presented on the screen, what she hears in speeches and chants, and the stream of comments from those tuning in from digital devices. Live streams are also often accessible after they finish, and the opportunity to rewatch segments blends the synchronous and asynchronous features of political action in "real life" as opposed to online. It is therefore a different type of listening, as the researcher can revisit recordings at any time; the opportunities to rewind, replay at different speeds, and toggle to specific seconds within a clip can empower researchers to engage more thoroughly with the events they study.

For researchers working in a non-native language, the opportunity to rewatch segments might allow them to hear phrases they may have missed, thus allowing for deeper engagement and understanding of the events themselves. I often did this, taking a few minutes to translate and double-check that I understood all the words properly after getting a text from an informant, whereas in a face-to-face interview, I may have just tried to guess the meaning from the context of our conversation or interrupted the participant to clarify a word's meaning. I can also google the name of key figures, like politician and speakers, in order to understand the references immediately, rather than furtively searching on my phone under the table to join in the conversation before the moment passes. The work of chatting via text or voice message is more disjointed than an in-person conversation; I can think carefully about my word choice and can delete a message to rephrase my question if I said "umm" too many times or made a grammatical mistake.

Of course, like all methods, it is important to acknowledge that there are limitations to digital ethnography. As I experienced, digital ethnography is not quite the same as being "in" the field, and reliance on the method risks overstating the importance of virtual interactions vis-à-vis "real" ones. Another challenge is that this method is vulnerable to sampling bias; the accounts I follow are neither exhaustive nor representative (Gray 2016, 503) because they privilege the perspectives of those who have access to the Internet.[4] However, a wealth of social science research – both in the sense of studying how people experience social and political phenomena online (Boellstorff 2008; Bonilla and Rosa 2015) and in using digital modes of access and analysis (Postill and Pink 2012; Sanjek and Tratner 2015; Gray 2016; Howlett 2021) – challenges arguments that suggest digital ethnography relies on a neat cleavage between online and offline behavior. Rather, digital and analog forms of engagement, community-building, and mobilization are "mutually co-constitutive" (Juris 2012, 260). Interpretive ethnography, as a method that prioritizes deep reflection about the process of data collection and analysis, is thus better suited to account for silences, the work of power in research relationships, and the meanings people ascribe to political and social phenomena.

Another concern about digital ethnography stems from the question of compounding inequalities around access to information. Relying on social media platforms as sites for research and a source of data risks further marginalizing those who already lack access to digital spaces. This concern strikes me as pernicious, insofar as it recalls the third feature of the "discourse of danger" that pervades thinking and writing on Central Asia, particularly the framing of the region as oriental. Orientalism is about more than situating Central Asia in some vague and borderless "East," as the concept's genealogy has roots in colonialist discourses that imagine non-Western cultures as exotic, backward, and uncivilized – all of which make the Orient appear dangerous (Said 1979). I thus worry about assuming widespread imbalances in connectivity because such an assumption threatens to perpetuate discourses of Central Asia's isolation and underdeveloped telecommunications infrastructure. Miller and Slater (2000) make a similar point about assumptions around technological backwardness in their study of digital culture in Trinidad at the turn of the century, and Moran (2015) goes further to complicate the normative underpinnings of the digital divide. She challenges assumptions that those on the "modern" side are more advantaged by reflecting on the infrastructure that allowed her to continue fieldwork in Liberia remotely. Creative workarounds to expensive infrastructure enabled Moran's interlocutors to call her much more cheaply than if she had dialed a Liberian number. Both studies accordingly offer evidence that published figures and prevailing imaginations of developing countries can very much underrepresent local populations' telephone and Internet access.

Since my first visit to Kyrgyzstan in 2014, I have been without Internet access only a few times. Specifically, I was offline during the first four weeks of my Peace Corps service when we were prevented from buying SIM cards, the few times I ran out of mobile credit while I was on a cheap phone plan in mid-2015, when I forgot my Beeline SIM card on visits to villages in the north where the Oshka SIM used in southern Kyrgyzstan provided weak service, and at a campsite high-up in the mountains of Jalalabad province. When I returned to Kyrgyzstan in 2018 after my first year of graduate school, it was even easier to stay connected because the phone companies were now offering unlimited data a month for approximately 3 USD, almost exactly the same amount I had paid for two gigabytes three years before. This was not just my experience as a privileged foreigner who could pay for a phone plan, though. The places I lived in and traveled to throughout the country represented a wide range of communities across the urban-rural spectrum, and my host families and colleagues held vastly different socioeconomic statuses but managed to stay connected to the Internet across these contexts. Even in the houses that did not have indoor plumbing, most family members still had their own smartphones. Those who did not often borrowed someone else's in order to toggle between several social media accounts to keep in touch with me 11 time zones away. My anecdotal experiences of broad connectivity in Central Asia are reflected by World Bank data, which finds that Internet access in Kyrgyzstan has skyrocketed from 20 percent in 2012 to 51 percent in 2022 (World Bank 2022).[5]

Just as I learned to listen against dominant theories about nationalism and instability in Kyrgyzstan, the work of translating my doctoral dissertation to a project about contentious politics based largely on digital ethnography and listening from halfway across the world created another opportunity for deep listening. Given this opportunity to observe how politics and governance straddle physical and digital spaces, I donned my ethnographer's hat and settled in to listen. Rather than defaulting to discourses of inaccessibility and digital illiteracy, and thus discounting findings from digital ethnographic studies, I came to realize the importance of adopting a flexible and curious outlook to explore how inequality in Internet access might map onto social cleavages, how self-presentation online resembles or deviates from in-person performance of identity, and how different people ascribe meaning to community-building, mobilizing, and claims-making both online and offline. As I learned through my fieldwork, scholars should listen widely, deeply, comparatively, and reflexively with the goal of understanding how forms of communication and political behavior work both online and offline.

Conclusions: A Call for Deeper Listening in the Discipline

In this chapter, I reflected on the process of curating and practicing a range of listening practices through three discrete phases of fieldwork. As a Peace Corps volunteer in Kyrgyzstan, I learned (and later had to unlearn) to be hypervigilant about the risks to my personal and national security while in the field, always listening for hints of impending danger in order to guard myself from violence through immersion and local language skills. In the early years of graduate school, as I was being disciplined within the field of political science, I realized the importance of listening against the grain of taken-for-granted academic concepts. The work of interrogating dominant academic theories and discourses calls for reflexive listening, not only in the sense of humbly considering what informants choose to share with us but also through continuous attention to time as a context that shapes our understanding of the world. Finally, tracing my experiences in coming to terms with being grounded in a physical sense due to the COVID-19 pandemic, I laid out tentative strengths and weaknesses of transitioning to digital ethnographic research which calls for me to listen from a distance.

Notably, the COVID-19 pandemic forced many PhD dissertation writers and early-career scholars to rethink their research agendas, but at the macro-level, it also represents a critical juncture that has laid bare deep inequalities within academia broadly and Central Asian studies specifically. As the academic world turned to video conferencing for teaching, roundtables, and conferences, Central Asian studies in particular flourished. While Eurasianists gather once or twice a year at region-specific conferences – such as the annual meetings of the Central Eurasian Studies Society or the Association for the Study of Nationalities – in 2020, there was a proliferation of events with hundreds in the virtual audience to listen to presentations on politics, history, medicine, and security in Central Asia. Organizers often took time zones into consideration, scheduling events through American and European universities to ensure that peers calling in from Tashkent or Almaty could join. Some panels equally took advantage of the possibility for simultaneous translation or the addition of subtitles to recordings of panels to make English-language discussions more accessible to Russian-speaking scholars. Presenters and audience members also did not have to go through the expensive, invasive process of applying for visas in order to travel abroad to get access to the information shared through these events.

These are meaningful and impactful changes that I hope will not fade away as vaccines roll out and borders slowly reopen. With this unique opportunity to reshape our epistemic and academic communities, it is imperative to think beyond measures to offer more voices a seat at the table. Some feminist theorists emphasize the importance of "giving voice" to marginalized groups and listening to subordinated others (for example, Hyams 2004), but listening to those who have been *given* voice out of good will still reflect a stark power dynamic. What good is a seat at a table that is structurally unsound, heavy and lopsided with economic, social, linguistic, gendered, and political inequalities? Is the answer to shove a book under the table's wobbly legs to balance it, or to flip the whole thing over, lay down *tushuk*s, and talk for hours over tea and treats? Who should we listen to when it comes to addressing the deep inequalities of the field – not in the sense of ethically researching some place "over there," but the epistemic community in which we are trained as social scientists and regional experts?

It should come as no surprise that women from Central Asia have been thinking about these questions for a long time, articulating the unique challenges of navigating layered inequalities at home and in the global academic market (Du Boulay 2019; Mamadshoeva 2019). Ethnographic methods and fieldwork have traditionally been based on the premise of a researcher traveling to a foreign context and struggling with the social realities of being an outsider, but, ultimately, benefiting socially and professionally from this perspective on the margins. Even with a critical eye toward the dynamics between insider and outsider, though, it seems that foreign scholars are not always the outsiders they understand themselves to be (Thibault 2021). Laura Adams, for example, describes the symbiotic relationship between the "mascot" researcher and the host; she notes that the status bestowed by nationality and profession often creates opportunities and incentives for hosts to push the researcher to the center (1999,

341–342). Ironically, scholars from Central Asia – especially women – can often be outsiders in their home societies (Suyarkulova 2019). Women scholars from the West are not only in a position to enjoy subjectively different insider status than their colleagues who were born, raised, and educated in Central Asia, but they also benefit from what Sultanalieva (2019) describes as the "coloniality of knowledge production." Central Asians "are the source material, the 'field,' the very fuel that feeds the production of knowledge about us but not for us."

Suyarkulova's and Sultanalieva's essays speak to a central part of Heathershaw and Megoran's definition of the "discourse of danger": that any danger posed by the field is centered on the experiences and anxieties of Westerners "while failing to appreciate the insecurities that are felt and experienced by Central Asians" (Heathershaw and Megoran 2011, 594). I worry that any concrete advice for fieldwork I can offer based on the five years I have spent in Kyrgyzstan would replicate the discourse of danger. The randomness of violence from state security forces, sexual predators, and visa regimes means that advice to take minibuses rather than taxis cannot protect you any more than if I advised the opposite. Such practical advice reinforces the primacy of this discourse of danger in field guides for women who research the region. Instead, I can point to the three pillars of ethnographic research – reflexivity, positionality, and the elucidation of meaning – as calling for patient but counterbalanced listening following from humility and respect.

Notes

1 This research was approved by Columbia University's Institutional Review Board. Conditions attached to the approval included obtaining verbal consent from interviewees, anonymizing the identities of teachers and mid-level bureaucrats (those who do not work at the national level), and securing physical and digital field notes.

2 Kyrgyzstan is one of the world's most ethnically diverse countries. More than 80 ethnic groups live in Kyrgyzstan, 11 of which have populations greater than 20,000, according to Kyrgyzstan's National Statistics Committee (2020). In the sphere of education, five ethnolinguistic groups are particularly salient: Kyrgyz (73 percent of the population), Uzbek (15 percent), Russian (5 percent), Tajik (1 percent), and Dungan (1 percent). State-sponsored curricula exist in Russian, Kyrgyz, Uzbek, and Tajik, and a new initiative offers Dungan language courses in some districts of the country.

3 Reeves (2014), Megoran (2017), and Liu (2012) are strong examples of long-term qualitative research conducted by foreigners in Kyrgyzstan.

4 In the sections below on limitations and possibilities of digital ethnography, I am indebted to conversations I had (over Zoom, nonetheless) at roundtables organized by the Ethnography Collective at the University of Massachusetts Amherst and the Interpretive Methods Clinics organized by Dvora Yanow and Peregrine Schwartz-Shea.

5 Together with Jasmin Dall'Agnola, I contextualize Internet use in Kyrgyzstan compared to other countries in Central Asia in our article "The Mobilizing Potential of Communication Networks in Central Asia" published in 2022 at *Central Asian Affairs*.

References

Ackerly, B. and J. True. 2008. "Reflexivity in Practice: Power and Ethics in Feminist Research on International Relations." *International Studies Review* 10 (4): 693–707. DOI: 10.1111/j.1468-2486.2008.00826.x.

Adams, L. 1999. "The Mascot Researcher: Identity, Power, and Knowledge in Fieldwork." *Journal of Contemporary Ethnography* 28 (4): 331–363. DOI: 10.1177/089124199129023479.

Al-Hindi, K.F. and H. Kawabata. 2002. "Toward a more fully reflexive feminist geography." In *Feminist Geography in Practice: Research and Methods*, edited by P. Moss, 103–116. Wiley-Blackwell.

Allina-Pisano, J. 2009. "How to Tell an Axe Murderer: An Essay on Ethnography, Truth, and Lies." In *Political Ethnography: What Immersion Contributes to the Study of Power*, edited by E. Schatz, 53–73. Chicago and London: The University of Chicago Press. DOI: 10.7208/9780226736785-005.

Anceschi, L. 2015. "The persistence of media control under consolidated authoritarianism: containing Kazakhstan's digital media." *Demokratizatsiya: The Journal of Post-Soviet Democratization* 23 (3): 277–295.

Are, C. 2020. "How Instagram's Algorithm Is Censoring Women and Vulnerable Users but Helping Online Abusers." *Feminist Media Studies* 20 (5): 741–744. DOI: 10.1080/14680777.2020.1783805.

Behl, N. 2017. "Diasporic Researcher: An Autoethnographic Analysis of Gender and Race in Political Science." *Politics, Groups, and Identities* 5 (4): 580–598. DOI: 10.1080/21565503.2016.1141104.

Beissinger, M.R. 2017. "'Conventional' and 'Virtual' Civil Societies in Autocratic Regimes." *Comparative Politics* 49 (3): 351–371. DOI: 10.5129/001041517820934267.

Boellstorff, T. 2008. *Coming of Age in Second Life: An Anthropologist Explores the Virtually Human*. Princeton: Princeton University Press. DOI: 10.2307/j.ctvc77h1s.

Bonilla, Y. and J. Rosa. 2015. "#Ferguson: Digital Protest, Hashtag Ethnography, and the Racial Politics of Social Media in the United States." *American Ethnologist* 42 (1): 4–17. DOI: 10.1111/amet.12112.

Dall'Agnola, J. and C. Wood. 2022. "The Mobilizing Potential of Communication Networks in Central Asia." *Central Asian Affairs* 9: 115. DOI: 10.30965/22142290-12340013.

de Seta, G. 2020. "Three Lies of Digital Ethnography." *Journal of Digital Social Research* 2 (1): 77–97. DOI: 10.33621/jdsr.v2i1.24.

Du Boulay, S. 2019. "The Moral Education of a Young Woman in Kazakhstan." *openDemocracy*, December 20. https://www.opendemocracy.net/en/odr/moral-education-young-woman-kazakhstan/.

Emerson, R., R. Fretz and L. Shaw. 2011. *Writing Ethnographic Fieldnotes*, 2nd edition. Chicago: University of Chicago Press. DOI: 10.7208/chicago/9780226206868.001.0001.

Eshaliyeva, K. 2020. "Why Domestic Violence Is Flourishing in Kyrgyzstan – and How It Could Stop." *openDemocracy*, February 20. https://www.opendemocracy.net/en/odr/why-domestic-violence-flourishing-kyrgyzstan-and-how-it-could-stop/.

Eurasianet. 2013. "Kyrgyzstan: Uzbek-Language Schools Disappearing." *Eurasianet*, March 6. https://eurasianet.org/kyrgyzstan-uzbek-language-schools-disappearing.

Fujii, L. 2010. "Shades of Truth and Lies: Interpreting Testimonies of War and Violence." *Journal of Peace Research* 47 (2): 231–241. DOI: 10.1177/0022343309353097.

Fujii, L. 2017. *Interviewing in Social Science Research: A Relational Approach*. New York: Routledge.

Gray, P. 2016. "Memory, Body, and the Online Researcher: Following Russian Street Demonstrations via Social Media." *American Ethnologist* 43 (3): 500–510. DOI: 10.1111/amet.12342.

Heathershaw, J. and N. Megoran. 2011. "Contesting Danger: A New Agenda for Policy and Scholarship on Central Asia." *International Affairs* 87 (3): 589–612. DOI: 10.1111/j.1468-2346.2011.00992.x.

Howlett, M. 2021. "Looking at the 'Field' through a Zoom Lens: Methodological Reflections on Conducting Online Research during a Global Pandemic." *Qualitative Research* (January): 1–16. DOI: 10.1177/1468794120985691.

Hyams, M. 2004. "Hearing Girls' Silences: Thoughts on the Politics and Practices of a Feminist Method of Group Discussion." *Gender, Place & Culture* 11 (1): 105–119. DOI: 10.1080/0966369042000188576.

Johansson, L. 2015. "Dangerous Liaisons: Risk, Positionality and Power in Women's Anthropological Fieldwork." *Journal of the Anthropological Society of Oxford* 7 (1): 55–63.

Juris, J. 2012. "Reflections on #Occupy Everywhere: Social Media, Public Space, and Emerging Logics of Aggregation." *American Ethnologist* 39 (2): 259–279. DOI: 10.1111/j.1548-1425.2012.01362.x.

Kapiszewski, D., L. MacLean and B. Read. 2015. *Field Research in Political Science: Practices and Principles*. Cambridge: Cambridge University Press. DOI: 10.1017/CBO9780511794551.

Koch, N. 2020. "Deep Listening: Practicing Intellectual Humility in Geographic Fieldwork." *Geographical Review* 110 (1–2): 52–64. DOI: 10.1111/gere.12334.

Liu, M. 2012. *Under Solomon's Throne: Uzbek Visions of Renewal in Osh*. Pittsburgh: University of Pittsburgh Press. DOI: 10.2307/j.ctt5vkgj5.

Mamadshoeva, D. 2019. "Listening to Women's Stories: The Ambivalent Role of Feminist Research in Central Asia." *openDemocracy*, October 9. https://www.opendemocracy.net/en/odr/listening-to-womens-stories-the-ambivalent-role-of-feminist-research-in-central-asia/.

Margolis, H. 2015. "'Call Me When He Tries to Kill You.' State Response to Domestic Violence in Kyrgyzstan." *Human Rights Watch*, October 28. https://www.hrw.org/report/2015/10/28/call-me-when-he-tries-kill-you/state-response-domestic-violence-kyrgyzstan#_ftn58.

Megoran, N. 2017. *Nationalism in Central Asia: A Biography of the Uzbekistan-Kyrgyzstan Boundary*. Pittsburgh: University of Pittsburgh Press. DOI: 10.2307/j.ctt1vjqrk6.

Miller, D. and D. Slater. 2000. *The Internet: An Ethnographic Approach*. Oxford: Berg. DOI: 10.5040/9781474215701.

Moran, M. 2015. "The Digital Divide Revisited: Local and Global Manifestations." In *eFieldnotes: The Makings of Anthropology in the Digital World*, edited by R. Sanjek and S. Tratner, 65–78. Philadelphia: University of Pennsylvania Press. DOI: 10.9783/9780812292213.

Moss, P. 2002. *Feminist Geography in Practice: Research and Methods*. London: Wiley-Blackwell.

National Committee on Statistics of Kyrgyzstan. 2020. "Ofitsiyal'naya Statistika: Naselenie." Available at: http://www.stat.kg/ru/statistics/naselenie/ (accessed on April 29, 2021).

Nilan, P. 2002. "Dangerous Fieldwork Re-Examined: The Question of Researcher Subject Position." *Qualitative Research* 2 (3): 363–386. DOI: 10.1177/146879410200200305.

Ortbals, C. and M. Rincker. 2009. "Fieldwork, Identities, and Intersectionality: Negotiating Gender, Race, Class, Religion, Nationality, and Age in the Research Field Abroad." *PS, Political Science & Politics* 42 (2): 287–290. DOI: 10.1017/S104909650909057X.

Peace Corps Office of Inspector General. 2019. "Final Country Program Evaluation Peace Corps/Kyrgyz Republic." Available at: https://s3.amazonaws.com/files.peacecorps.gov/documents/inspector-general/Final_Report_on_the_Evaluation_of_Peace_Corps_Kyrgyz_Republic_IG-19-08-E.pdf (accessed on December 1, 2020).

Postill, J. and S. Pink. 2012. "Social Media Ethnography: The Digital Researcher in a Messy Web." *Media International Australia* 145 (1): 123–134. DOI: 10.1177/1329878X1214500114.

Reeves, M. 2014. *Border Work: Spatial Lives of the State in Rural Central Asia.* Cornell: Cornell University Press. DOI: 10.7591/9780801470899.

Said, E. 1979. *Orientalism.* New York: Vintage.

Sanjek, R. and S. Tratner. 2015. *eFieldnotes: The Makings of Anthropology in the Digital World.* Pennsylvania: University of Pennsylvania Press. DOI: 10.9783/9780812292213.

Schrooten, M. 2015. "Writing eFieldnotes: Some Ethical Considerations." In *eFieldnotes: The Makings of Anthropology in the Digital World*, edited by R. Sanjek and S. Tratner, 78–93. Philadelphia: University of Pennsylvania Press. DOI: 10.9783/9780812292213-006.

Sirnate, V. 2014. "Positionality, Personal Insecurity, and Female Empathy in Security Studies Research." *PS, Political Science & Politics* 47 (2): 398–401. DOI: 10.1017/S1049096514000286.

Sultanalieva, S. 2019. "How Does It Feel to Be Studied? A Central Asian Perspective." *openDemocracy*, October 8. https://www.opendemocracy.net/en/odr/how-does-it-feel-be-studied-central-asian-perspective/.

Suyarkulova, M. 2019. "A View from the Margins: Alienation and Accountability in Central Asian Studies." *openDemocracy*, October 10. https://www.opendemocracy.net/en/odr/view-margins-alienation-and-accountability-central-asian-studies/.

Thibault, H. 2021. "'Are You Married?': Gender and Faith in Political Ethnographic Research." *Journal of Contemporary Ethnography* 50 (3): 395–416. DOI: 10.1177/0891241620986852.

Tufekci, Z. 2017. *Twitter and Tear Gas: The Power and Fragility of Networked Protest.* Yale: Yale University Press. DOI: 10.12987/9780300228175.

Veletsianos, G., H. Shandell, H. Jaigris and G. Chandell. 2018. "'Women Scholars' Experiences with Online Harassment and Abuse: Self-Protection, Resistance, Acceptance, and Self-Blame." *New Media & Society* 20 (12): 4689–4708. DOI: 10.1177/1461444818781324.

World Bank. 2019. "Individuals Using the Internet (% of Population) –Kyrgyz Republic." https://data.worldbank.org/indicator/IT.NET.USER.ZS?locations=KG.

7 The Academic Lion Skin

Balancing Doctoral Research with Motherhood

Ruta Skriptaite
University of Nottingham

Introduction

I was three-months pregnant when I was first interviewed for my doctoral position in July 2018. I was determined to study masculinity as it is a key aspect of the political leadership and imagery in the post-Soviet space, a region where I was born. Growing up in a predominately patriarchal post-Soviet Lithuanian society, where members of the government are still commonly referred to as the *men of the government*,[1] I always felt compelled to question the patriarchal order of things. The subject of my doctoral research project "Political Image-making and Post-Soviet Patriarchal Leadership: a comparative analysis of Belarus, Kazakhstan and Turkmenistan" is informed by three factors. The first two were my fascination with political image-making as well as the power of visuals and the lack of academic literature on political image-making as it relates to the three case studies. As Roland Bleiker put it, "[t]here is something unique about images. They have a special status. They generate excitement and anxieties." (2015, 875) Political image-making, as it relates to my research, can be defined as the strategic manipulation of their appearance and image by political leaders to maximize their appeal to voters. This manipulation is quite often carried out through imagery employing different media (e.g., photos, TV, print) but can also be channeled through various visual means, including monuments depicting leaders, and buildings or certain locations, such as cities, being named after them. Yet, political image-making is not solely confined to the visual realm. It can be constructed using both visual and verbal messages (e.g., speeches, tweets) (Lalancette and Raynauld 2019). Within the post-Soviet region, political leaders are known for relying on projections of gendered power (e.g., hypermasculinity) as a strategy for creating not just legitimacy but also maintaining their power (Sperling 2015; Ashwin and Utrata 2020). For instance, Alyaksandr Lukashenka and Vladimir Putin, when dressing in ice-hockey gear or riding bare-chested on horseback, rely heavily on heteronormative ideas of masculinity – in this case, performances of almost cartoonish masculine bravado – to appeal to the broader public and especially their voter base. While much has been written about Putin's political

DOI: 10.4324/9781003144168-11

image, the strategies employed by other authoritarian leaders in the region to manipulate their political images have gained less attention. Therefore, I have chosen the political images of the current Belarusian leader Alyaksandr Lukashenka, the former Kazakh president Nursultan Nazarbayev and the deceased Turkmen leader Saparmurat Niyazov-Turkmenbashy as the primary objects of my research, especially as it relates to the affinity between traditional masculinity and political leadership.

Whilst my initial choice of research project was primarily based on a fascination with political imagery as well as the under-researched nature of the topic, the final and third pillar on which my PhD project was built can be spelled out using the below mentioned quote by Carole Pateman about the political lion skin so ominously unfitting to females:

> The political lion skin has a large mane and belonged to a male lion, it is a costume for men. When women finally win the right to don the lion skin it is exceedingly ill-fitting and therefore unbecoming.
>
> (Pateman 1995, 6)

According to Pateman, the political lion skin fits women so poorly because it has been designed throughout human history to accommodate men. Similarly, my experience as a working-class female (who was raised by a single mother), a foreigner in the country where I reside and am building my career, and by status as a parent – in addition to being a PhD student – are factors that make my academic lion skin "ill-fitting" or at least requiring constant alterations.[2]

While this is true in any context, it was most notably observed during the COVID-19 pandemic as I conducted online interviews for my doctoral research. This chapter therefore highlights the ways identities and positionality, especially as both a PhD researcher and a working-class mother, impacted my academic trajectory. By drawing on my own experiences, my research touches upon the path that led me to become a doctoral researcher, the impact of the global pandemic on my research methods and strategies, as well as the challenges and dilemmas I experienced while carrying out online interviews. The following pages also highlight the pandemic's gendered side, especially the unique challenges faced by mothers working in academia.

This chapter is divided into three sections. In the first, I describe my path to academia through a gendered lens – while sharing my experiences as a female applicant for a doctoral degree with parental responsibilities and touching upon other aspects of my background: being a foreigner in the country I chose to pursue my career in, who herself is from a working-class background. From there I reflect on my online recruitment of interviewees – and discuss the obstacles I encountered. Afterwards, I discuss how the global pandemic has influenced my research, drawing upon the abovementioned narrative of a young female researcher with parental responsibilities.

I outline my experience of conducting online interviews for the purpose of researching political image-making and gender as it relates to my three case studies – Belarus, Kazakhstan, and Turkmenistan – three countries that are widely considered to be patriarchal and authoritarian.

Contingencies, My Pregnant Body, and the Journey to Academia

Women academics are not a single entity, hence they do not all experience the same disadvantages. It is very much dependent upon intersectionality – the other parts of their identity – be it their race and ethnicity, class, or any other factors that are considered to be a part of their identity. I do not object to the statement that, overall, females are disadvantaged in comparison to males when it comes to career-building, including in the academic sector. However, the circumstances mean that the lion skin fits even worse on some. I would like to discuss the disadvantages by drawing upon my own experience.

Ironically enough, academia, despite its primary purpose of being at the service of the advancement of humankind, still lags behind when it comes to gender equality. For instance, women scholars publish fewer articles, are awarded fewer grants, and receive fewer citations, and their chances of being promoted or granted tenure in comparison to their male counterparts are lower (Hechtman et al. 2018; Catalyst 2020; Huang et al. 2020 in Deryugina, Shurchkov and Stearns 2021, 2). Thus, the sole fact of being a female means you are not seen as a likely candidate for a successful academic career. For instance, the year that I started my PhD study program, there were four other female researchers starting their degrees at the School of Politics and International Relations at the University of Nottingham. In contrast, 13 males enrolled at the same time. Indeed, it is common knowledge that in Higher Education it is still quite often the case that a variety of factors constituting an individual's identity, such as gender, sexuality, or race and their intersections, have an impact on their career opportunities. This is particularly evident for those who do not embody the stereotype of "a typical" academic – a white male (Wright, Haastrup and Guerrina 2021, 164).

In addition, my life experience made me aware of the interconnectivity of these disadvantages. I was better able to understand the disadvantages I faced as a female with parental responsibilities who originates from the post-Soviet space, as well as from a relatively disadvantaged background (I grew up in a single-parent family). I have experienced harassment and negative comments linked to my ethnicity as a result of the prejudices toward people from the post-Soviet space. For instance, on one occasion, not long after finishing my undergraduate degree at Essex University, a fellow student told me that my career path would not progress and that I would end up cleaning tables at Costa, because "this is what Lithuanians do when they come to the UK." This was the moment I realized that women's empowerment and gender equality is an issue related to one's class and geographical origins. Due to my socioeconomic background, even whilst pursuing my master's degree, I often had to work evening shifts in catering to make a living in London. In fact, even though the MA program that I was enrolled in at the School of Slavonic and Eastern European Studies (UCL) would normally have taken a year to complete, due to my financial circumstances I had to do it part-time and also take a gap year between the two years of my part-time studies so that I could gather the funds for the second half of my degree. Naturally, this meant that in contrast to my classmates, who usually met up to socialize and debate about politics off-campus, I quickly packed my books and rushed off to waitress at another event. Indeed, while completing my undergraduate degree, I would question whether the fact that instead of attending protests in London I was spending my extracurricular hours doing low-paid agency jobs would make me apolitical as a student of political science. Despite feeling like an alien due to my class background, shouldering the double burden of studies and part-time work taught me a lot about time management. Yet, I was again reminded of my class background as a future researcher a few years later – when I was getting ready for my journey as a PhD.

From the very beginning, my experience of academia went hand in hand with my working-class background and parenthood. Whilst still a PhD applicant (summer of 2018 to January 2019) I was working in a customer-facing role – as a receptionist and administrator in central London. This was around the time when I found out that I was pregnant with my first daughter and decided to pursue both life-changing opportunities simultaneously. My circumstances at the time were not ideal, at least in conventional terms, for starting a family. My partner and I still lived in different cities in the UK, and I finally had a real chance to start my long-awaited PhD program. Interestingly, and also quite frustratingly, as my pregnancy became more evident, I began getting unwanted attention related to it. My decision to choose both parenthood and a career in academia was met with a range of reactions and emotions from my co-workers and friends, which ranged from disapproval to admiration. While trying to keep my situation private, I was unable to escape the various gendered reactions that related to my choice and circumstances. For instance, opinions such as the following were often expressed to me: "a woman does not need a PhD," "you do not need a baby, just get on with your PhD plans," "you cannot afford both." I, as a feminist, found it quite difficult to cope with the fact that people I did not usually engage with suddenly felt the need to discuss my pregnancy with me, suggesting that my entire life now revolves around the pregnancy. Some even touched my "baby bump" without my prior consent. Indeed, as Beech, Kaufmann, and Anderson (2020, 523) state, in some instances, I felt as if I was seen as merely a "human incubator" and that my pregnant body "is treated as public property". Experiences like these, as I reflect on them over three years later, presented an uncomfortable conflict between how I wanted to be seen and how I was considered by others, at least in my view. Indeed, my pregnant body, or rather my co-workers' gaze directed at it, in some instances consequently made me feel like an outcast and already excluded from a space I was yet to enter.

This leads me to one of the major issues, that I would like to address in this chapter – the so-called motherhood penalty that women, who choose to have a family, often endure. The motherhood penalty can be shortly defined as a disadvantage experienced by women who want to have both a career and children. These disadvantages are twofold. First, the female biological makeup plays a role when it comes to personal and career-life balance, since women are the childbearers. When choosing to give birth to a child, women are expected to take at least a minimal break from their career, irrespective of how career-driven they are. The physically demanding process of childbirth and the infant's dependence on its mother, if she chooses to breastfeed, are a given. Because of their biology, women must take on several roles. Some of those outlined above will affect their career trajectory. Second, traditional (and archaic) social discourses around child-rearing and housekeeping place women at another disadvantage as they are regarded as homemakers by default. Budig and England (2001 in Pepping and Maniam 2020, 113) suggest that childcare and domestic chores affect a woman's performance in the labor market as well as her career prospects. This rededication of available time can also result in mothers being perceived as less competent or committed to their job and career.

In contrast to my private life, though, I experienced the absence of prejudice against a pregnant woman's body whilst attending interviews as a PhD applicant. I attended three face-to-face interviews when applying for my PhD. Two of them took place in the summer of 2018 – during the early stages of my pregnancy. The third one – at the beginning of 2019 – approximately a month prior to giving birth to my first daughter. I recall being extremely self-conscious during the interview and hoping that the two interviewers, among whom was my potential supervisor, would (somehow) not notice that I was heavily pregnant. To my relief, no one brought it up during the meeting. However, after the interview, the academic, who could have been my supervisor (I ended up accepting an offer from a different university), invited me to a coffee shop. There, whilst having an informal conversation, she brought up my pregnancy, also sharing with me that she has a small child of her own. She advised me to dedicate at least a couple of days per week just to my research. The conversation provided me with a sense of belonging – it made me feel less of an alien or imposter in the world of academia. After countless applications to universities and various sources of funding, my efforts were rewarded. I was offered partial funding (a fee waiver) by the University of Nottingham, meaning that against the odds I was about to start my long journey as a doctoral researcher. I began the program at the end of September 2019. At that time, my first daughter was six months and two weeks old.

Studying Authoritarianism during the Pandemic: Virtual Snowballing and Digital Surveillance

The choice of my case studies – the three leaders of the post-Soviet space: Alyaksandr Lukashenka, Nursultan Nazarbayev, and Saparmurat Niyazov-Turkmenbashi – was primarily based on my fascination with their political images and image-making techniques as well as the under-researched nature of the topic. Indeed, the three leaders can be considered as arguably some of the most authoritarian and patriarchal presidents in the post-Soviet space yet also relatively underexplored as regards political image-making and especially its gendered aspects. They emerged as the fathers of their newly created states in the early 1990s – a time of the rise and fall of many patriarchal regimes. The "fathering effort" of the three patriarchs – masculinity-based style and methods of political image-making, although archaic and outdated – still has a remarkable impact and hinders the achievement of equality in the societies they reign(ed) over for decades. For instance, Lukashenka is well known for his infamous statement that it is "better to be a dictator than a gay person" (Reuters 2012) and dismissing the three women, Sviatlana Tsikhanouskaya, Mariya Kalesnikava, and Veranika Tsepkala, who stood up to him during the 2020 presidential ballot and created the United Front – a pro-democracy movement challenging the authoritarian Lukashenka's reign – as "poor things" implying women's unsuitability for political action (Sous 2021). Such a discourse indeed actively encourages the perpetuation of homophobia and misogyny. In the meanwhile, Nazarbayev is famous for his self-representation as the father of the independent Kazakh state, whilst the commemoration of this "fatherhood" could be seen across the country, starting with various monuments, such as the famous Bayterek tower, finishing with the renaming of the capital Astana to Nur-Sultan in his honor (The Guardian 2019). Last but not least, Niyazov-Turkmenbashy, in addition to building numerous monuments in his own honor, had dedicated his reign to "teaching" his nation the "proper way" of being Turkmen. His teachings were compiled and published as a book, "Rukhnama" (Turkmenbashy 1995), which still serves as a compulsory guide for people in Turkmenistan. For example, it is still part of the Turkmen school curriculum.

Despite growing political and social unrest in their countries, Alyaksandr Lukashenka and Nursultan Nazarbayev continue to influence local leadership practices and gender norms. Although he had been stripped of all his lifetime titles in the aftermath of the violent unrest in January 2022, Nazarbayev retains the title of *Elbasi* (the Kazakh word for Leader of the Nation) and the right to address parliament or, should he wish, attend government meetings on matters of major importance and even sessions of the constitutional council (Kumenov 2022). Like Nazarbayev, Lukashenka is still clinging onto power, despite his eroded legitimacy both inside and outside Belarus. Not even the March for Freedom – the largest recorded gathering in the history of Belarus (it is estimated that between 200,000 and 500,000 people took part) – was able to dethrone him (Belsat 2020, 2021). In contrast, Turkmenbashy's presidency, at least the end of it, took a significantly different path than those of the other two discussed leaders. He died whilst in office in 2006 (Pannier 2006). Nevertheless, his successor Gurbanguly Berdimuhamedov, who himself was recently succeeded by his own son Serdar Berdimuhamedov after the election of questionable fairness held on March 12, 2022 (Radio Free Europe Radio Liberty 2022), seems to be preserving his predecessor's tradition of self-glorification, gigantic golden statues, and patriarchy (Walker 2015).

When starting my course back in fall 2019, my initial plan was to conduct several overseas fieldwork trips in my three case study countries, Belarus, Kazakhstan, and Turkmenistan, during the summer break of 2020. First, those trips would have served as opportunities to visit numerous historical sites (e.g., statues, memorial monuments) and buildings (e.g., museums, residency palaces) important to the political image-making of the three presidents. In addition to visiting locations, I planned to conduct semi-structured interviews with female and male academics, journalists, political analysts, activists, and observers who were either from or had lived in the countries I studied for a significant length of time. Although all three of my case studies are generally considered to be authoritarian regimes, it is crucial to note that the levels of oppression in Turkmenistan are significantly higher than those in Belarus or Kazakhstan. Importantly, and due to these high levels of political oppression in Turkmenistan, I did not plan to carry out any interviews with individuals while physically present in the country. According to Human Rights Watch, "[the Turkmen] government severely restricts all fundamental rights and freedoms, including freedoms of association, expression and religion" (Human Rights Watch 2022). Indeed, back in 2019, when I was starting my research project, the country was considered significantly less free than Kazakhstan and Belarus. Reporters Without Borders placed Turkmenistan at the very bottom of their 2019 World Press Freedom Index list (180/180); in comparison, Belarus was placed at 153/180 and Kazakhstan at 158/180 (Reporters Without Borders 2019). My original plan was therefore to organize those interviews either online or outside of Turkmenistan to avoid any possible risk for my interlocutors. Given the state-led repression of civilians following the mass protest in Belarus that started in 2020, I decided to use a similar approach for my data collection in both Belarus and Kazakhstan, especially following the events commencing in January 2022.

In addition to the unstable political climate in my case study countries, changes to national and international spaces brought about by the ongoing pandemic resulted in having to significantly alter my research design. As Howlett (2021, 12) phrased it: the virus pushed me "back into [my] armchair – both in a physical and metaphorical sense." I had to shelve my plans to travel to the region and switch to online-based research. First, since the start of the pandemic my university has not been in favor of overseas fieldwork and travel in general. Also, given that most of my potential interviewees were based abroad, it made sense to move fully online.

Yet, switching to video-recorded, semi-structured interviews[3] on Microsoft Teams required many hours of administrative work amending my research ethics clearance. For instance, I had to declare that, for the foreseeable future, all my research related to gathering primary data will be done online and I will inform the University Research Ethics Committee if there is any change to the plan. In addition to the increased administrative burden, I was worried that my new research methods could risk lowering the quality of the work I would produce. Indeed, questions such as "Did my information-gathering techniques even qualify as fieldwork?" and "Did I deviate into the zone of 'armchair' research and lose or never establish the connection and interaction with the countries I was studying?" started to plague me whilst I was moving my research fully into the online space.

Despite my initial doubts about conducting my PhD research fully online, the saying "good things come when you least expect them" indeed sums up the recruitment process. In March 2021, I agreed to write an International Women's Day post for my school's blog on the gendered importance of Sviatlana Tsikhanouskaya's political leadership. Little did I know that my swiftly put-together piece titled "Political Lion Skin: what it means for Belarus to have a female opposition leader?" (Skriptaite 2021) would help me to recruit interview participants for my first case study, Belarus. About a month after the publication of my blog post, I checked the spam folder of my university's email account and came across a message from a university professor based in Belarus. The email sender was explaining that they had read my blog post and offered to put me in touch with valid sources of information on what is currently happening in Belarus. I responded apologizing for my delayed reply and asked whether they would be willing to be interviewed for my project. The professor agreed and provided me with the promised list of contacts. From there, my Belarusian interviews snowballed swiftly as one contact led to another.

While the strategy for recruiting interviewees for my first case study has worked very well, the interviewing for the last two case studies, Kazakhstan and Turkmenistan, is still in progress. Despite supplementing my participant list by asking people who have studied the area for leads and undertaking extensive online search for potential interviewees, as of August 2022, I still only had 16 interviews in my basket for Turkmenistan, in contrast to 25 for Kazakhstan, for which I started the recruitment process around the same time. I identified two reasons for the slow pace of progress in the Turkmenistan case study, including the limited number of people I could approach about my project, as well as the "red flag" nature of my research topic. What, in theory, seemed not such a "red flag" topic was, in practice, sometimes met with hesitation on the part of my potential interviewees. Given that my research topic explores the political leadership of regimes of an authoritarian nature, I came to understand that, to some of my participants, especially those based in the case study countries, being *out of line* with the official discourse, may have unwanted serious consequences (e.g., investigations by the police, detention) (Roberts 2012; Gentile 2013). All three regimes are known for using new high-tech tools, such as artificial intelligence, advanced biometrics, "smart filtering," and hacking spyware (e.g., Pegasus, Fisher), to manipulate and repress their citizens and political opponents on- and offline (Anceschi 2015; Polyakova and Meserole 2019; Radio Free Europe Radio Liberty 2019). In the case of Turkmenistan, Internet surveillance and the state's attempts to isolate the society from the international online space, for instance, by a crackdown on the use of VPNS (Radio Free Europe Radio Liberty 2021), is a big concern to Turkmens who do not agree with the regime. Their concerns seemed to be justified. The Turkmen government is known for using the UK FinFisher spyware, which is capable of remotely taking control of a computer or mobile phone and logging users' activities (Nazar 2018; Radio Free Europe Radio Liberty 2019, 2021). Those worries about the security of online activities and the Turkmen state's interference are felt even beyond the borders of the country. For instance, one of my Turkmen participants who was based abroad[4] told me that there have been concerns about attempted email hacking by the Turkmen state of the organization they are working for. So, it comes as no surprise that Turkmenistan is the only one of the case studies where potential participants declined my interview requests for reasons of safety. Moreover, while my respondents from Belarus and Kazakhstan usually agreed to sign the consent form, the procedure of providing written consent deterred a few of my Turkmen participants. Nevertheless, to address these safety concerns, I ensured that interviewees would remain fully anonymous and that collected data would be protected through all stages of the research process and even after the publication of the thesis. I also employed measures to accommodate the participants, such as the verbal consent option. I opted for verbal consent when a potential participant felt unsafe signing a written form.[5]

Despite the ever-changing schedule and design of the research, leading to plenty of administrative work and dialogues with the Research Ethics committee, I have now completed over 100 semi-structured interviews (in addition to my thesis, I am also currently interviewing for a collaborative project with one of my supervisors). However, my effort to find new participants is not over yet. In the last part of the chapter, I reflect on my experience of conducting interviews from my personal space – my home.

Online Interviews: When the Professional and Private Spaces Collide

The new contingencies brought about by the global pandemic made academics open to new ways of conducting their scholarly work (Sah, Singh and Sah 2020). Indeed, finding a contingency plan was crucial for researchers as they battled to meet deadlines in such uncertain times (Krause et al. 2021,

264–265). Despite the broad and multidimensional impact of the COVID-19 pandemic, individuals have faced different consequences, dependent on their gender, ethnicity, and class, as well as other identities (Kowal et al. 2020; Burzynska and Contreras 2020; Malisch et al. 2020). Various studies indeed exposed the existence of the academic lion skin. They showed that female researchers have been more disadvantaged than their male counterparts, with uneven childcare or housework distribution, for example (Burzynska and Contreras 2020; Malisch et al. 2020; Deryugina et al. 2021; Lerchenmüller et al. 2021). According to the study by Deryugina et al. (2021, 3) that featured a sample of 19,905 respondents with doctoral degrees (11,901 self-identified as men and 8,004 as women), even prior to the COVID-19 outbreak, female academics dedicated approximately 30 minutes less time per day to their research than their male counterparts. They also spent 40 more minutes on childcare duties than men. The research shows that this asymmetry became even more evident during the pandemic as, on average, women with parental responsibilities "lose about an hour of research time per day more than childless men do," whilst men with dependent children only lose 30 minutes of research time compared to men with no children (Deryugina et al. 2021, 7).

Despite the pandemic's asymmetric impact on academic mothers as well as my initial skepticism toward online interviews, my personal experience of interviewing online has been mostly positive. Firstly, conducting my research fully online helped me to overcome several barriers. Conducting research online during the pandemic was financially easier and helped me cope with parental responsibilities. Since my university awarded me only partial funding, I would have needed to use my own funds to do fieldwork abroad. Moreover, my parental responsibilities add further complications to traveling abroad. Carrying out my fieldwork abroad would have meant that I needed to travel with my whole family since both of my daughters are very young (the younger one turned one in April 2022). That would have also meant that my partner needs to take time off work to travel with me, so he could look after the children, for instance while I interview people. Thus, before the pandemic, I already faced the problem of not being able to visit all three post-Soviet states as scheduled. However, the shift to online working in academia made my research more affordable and manageable. I was no longer expected to travel abroad but allowed to do my fieldwork from home. This enabled me to optimize my time: instead of conducting interviews in only one post-Soviet country, I had the opportunity to interview multiple respondents in different locations on the same day. Moreover, online interviews also meant that I did not have to leave my children for a long period. For instance, at the present I only require childcare for a few hours a couple of days a week when I have interviews scheduled or need to be on the university campus teaching or attending to other duties as a PhD student. I found such arrangement manageable as I share childcare duties with my partner, and we also receive help from our families.

Yet, I cannot contrast my experience of online interviewing with that of face-to-face encounters, because by the time that I started recruiting my participants, the world was already immersed in the online space due to the global outbreak of COVID-19. However, that also provided me with an opportunity to interview individuals to whom I would have no access if I were to interview them in person. The geographical proximity-defying online space connected me with participants based in other continents, including North America and Australia, which I would not have been able to visit for my research even if there were no travel restrictions in place. Several social media and IT platforms like Facebook, Telegram, Twitter, and Microsoft Teams provided me with the means of locating and recruiting my participants as well as interacting with them without leaving my living space (Jowett, Peel and Shaw 2011; Deakin and Wakefield 2014; Sah et al. 2020, 1104–1105). Thus, my laptop has been my way into the field, which also serves as a barrier separating my private and public spheres as a researcher.

While, overall, I was happy with the outcome of my online interviews and the insights, thoughts, and information I was provided with, some interviews did leave me feeling like an alien in my field who is quite miserably failing to pose as a confident and competent researcher. For instance, one of my interviewees who happened to be senior to me in years and in terms of career status showed discontent with one of the initial questions on my list (indeed, not a sign of a great start to an interview). The situation would have probably been avoided if I, instead of dwelling on questions related to the concept of nationalism, which neither has direct relevance to my research topic nor is something I specialize in, would have moved to the core pillars of my research topic: political image-making and masculinity. Although feeling like an exposed imposter, there was indeed no way back – having two years invested in the actual studies for the PhD degree, plus countless years preparing for it – abandoning the project was not an option. Hence, I did exactly what many other researchers have done prior to me (or will do in the future for that matter) – scrapped a part of my questionnaire and came to accept that I am still in the process of learning and hopefully in the future will be able to tell myself that "this is something I am good at." Indeed, now after having carried over 100 interviews I learned that confidence and skills are something that come through work and experience, whilst gaining these two takes time and determination.

The international turbulence triggered by the spread of COVID-19 revealed what we see as "private" and "public" or "home" and "the field" as less than rigid and stable spaces (Till 2001).

I often found the nature of the barrier separating the "professional" from "private" unpredictable, as if by inviting my participants into my public space as a researcher for interviews, I was also inviting them to experience my personal sphere, which, in some instances, went beyond my interviewees just having a glimpse of the Clockwork Orange poster on the wall of the living room that doubles as my interview and meeting space. I have indeed witnessed the relativity of the meanings ascribed to "private" and "public" whilst conducting an online interview when my toddler was trying to climb on my lap, or when she knocked my laptop off my desk during a call with my supervisor. In those instances, the calls making me take the role of being observed rather than remaining in control as the observer. These moments of "exposure" and the inability to always control my private space and fully convert it into a professional one, at least immediately after the interview would make me feel like an imposter – a woman not fully in control of the domain traditionally assigned to her, who is miserably failing at posing as a professional – in a domain which by tradition is normally gifted to men. Such moments when private and public spaces merge would have not been experienced if my interviews had taken place offline.

Moreover, these moments would not have been experienced by an academic who is not so often doing her work in the presence of her children. Although I am lucky to have a partner to share childcare duties with and a family who supports me, in contrast to scholars who were born in the UK, I, as a migrant to the UK, could not always rely on my family and support networks because of the travel bans imposed by the pandemic. Indeed, I was lucky that my mother was stuck in the UK when the government announced the first lockdown and that she agreed to stay with us for several months during the third year of my PhD. Nevertheless, at a certain point my mother did have to leave due to her own work commitments. I must admit that conducting interviews and working in the presence of my children has put a lot of pressure on me and has had a significant impact on my productivity. Indeed, as an academic mother, I lack the extra time for reflection and research that working from home during a pandemic could potentially provide to scholars who do not have parental obligation. As Kowal et al. (2020, 390) rightly point out "it is difficult to picture Isaac Newton inventing calculus [during the Great Plague of London in 1665] while cooking dinner with a pre-schooler tugging on his sleeve, followed by teaching an online class."

Nevertheless, I did also experience the moments of reaffirming my rightful place in the world of academia. One of the most memorable ones was during an interview with a participant from Kazakhstan when we somehow ended up talking about children and she expressed her admiration for my raising two small daughters whilst pursuing a full-time PhD. Although, the conversation about the children was not elaborate enough to determine whether the positive attitude toward my maternal responsibilities stemmed from the glorification of motherhood in post-Soviet Central Asia (Kamp 2016) or the sense of solidarity in shared motherhood. Moments like these indeed moved me further away from the image of myself as someone who is just gambling away her family time and inspire hope that, one day, I will become a woman empowering her daughters by her own example.

The rocky road itself that I experienced during the journey pursuing my doctoral degree gifted me with resilience and determination – qualities that are crucial to building any successful career. Although disappointingly enough, partially due to the global pandemic and partially due to the political climate I might not be able to visit my case studies for a very long time, the unfortunate circumstances have also taught me that adaptability is the key to both success and survival. To me – on my journey as a researcher and as mother.

Conclusion

When starting my journey as a researcher in political science in September 2019, I was already aware of the problems that I would likely face, given my background and parental responsibilities. The global pandemic, which pushed the world into the space of virtual interactions and thinned the line between the domains of what we conventionally perceive to be "private" and "public," "personal" and "professional," added an extra layer to the already existing contingencies. My initial research plan that incorporated fieldwork in three post-Soviet states – Belarus, Kazakhstan, and Turkmenistan – was fundamentally altered as it had to be moved fully online, which also posed new risks around me being able to produce original work based on primary data. Nevertheless, the new situation created by the pandemic had aspects that favored my initial circumstances, especially the fact of me being a female researcher with parental responsibilities. These turned out to be helpful. Although denying me the normal access to my field, the pandemic, by opening the door to virtual space for academics, made my research more affordable and manageable. On the other hand, the merging of the private and professional domains, in addition to its benefits, also made me question my competence as a researcher when the two dimensions came too close. Whilst my journey to academia has always been marked by uncertainty, the time since the start of the pandemic made me appreciate the adaptability it offered, which I found to be a prerequisite for any successful project. And as for the lion skin, female researchers with parental responsibilities will carry on adjusting it into a better-fitting academic suit.

Notes

1 The term *valdžios vyrai [the men of government] is used not only in colloquial speech but also in formal language, as for instance in news articles* (LRT 1995; Bieliavska 2020; Mykolaityte 2020).

2 My eldest daughter is three years old and was born six months before the start of my doctoral studies, whilst I gave birth to the younger one during the second year of my PhD studies at the University of Nottingham.

3 In a few instances, my participants requested interviews in an audio call or as a written questionnaire. I understand that the main reasons for such requests were safety concerns to do with timing or wanting greater control of the outcome. Switching to online interviews required administrative work amending my research ethics clearance and participants' consent form.

4 Turkmenistan was also the only case study where a large majority of my participants are based abroad. As of August 2022, I have interviewed only one person who is based in Turkmenistan.

5 The university's Research Ethics Committee (REC) approved my suggested change in regard to participants' consent, under the condition that I would adhere to the general guidance of the REC as well as document each case of verbal/informal consent in my own register for the REC, specifying the date and place, the name of the interviewee, the purpose of the interview, and a probable reason why verbal/informal instead of written/formal consent has been obtained. In addition, I was asked to save my own register on the university's password protected OneDrive data storage space, which I also used to secure my collected data. Research shows that data storage is a key issue in the region which is the focus of my PhD project (Roberts 2012; Gentile 2013). In the privacy notice, I indicate measures taken to safeguard stored data, which includes the anonymization of data and password-protected online storage. So far, no participants have raised concerns regarding the way research data is stored.

References

Anceschi, L. 2015. "The Persistence of Media Control under Consolidated Authoritarianism: Containing Kazakhstan's Digital Media." *Demokratizatsiya: The Journal of Post-Soviet Democratization* 23 (3): 277–295.

Ashwin, S. and J. Utrata. 2020. "Masculinity Restored? Putin's Russia and Trump's America." *Contexts* 19 (2): 16–21. DOI: 10.1177/1536504220920189.

Belsat. 2020. "'Step Down', 'Stop Violence.' Biggest-Ever Protest Rally in Minsk." *Belsat*, August 16. https://belsat.eu/en/news/nationwide-march-for-freedom-vs-pro-lukashenka-rally-live/.

Belsat. 2021. "In Photos: Biggest-Ever Belarusian Street Protest Took Place Year Ago." *Belsat*, August 16. https://belsat.eu/en/news/16-08-2021-in-photos-biggest-ever-belarusian-street-protest-took-place-year-ago/.

Beech, O., L. Kaufmann and J. Anderson. 2020. "A Systematic Literature Review Exploring Objectification and Motherhood." *Psychology of Women Quarterly* 44 (4): 521–538. DOI: 10.1177/0361684320949810.

Bieliavska, J. 2020. "Valdžios vyrams teks atsakyti į klausimus dėl paauglėms skirtos kontraceptinės spiralės" Trans. [Men of the government will need to answer questions regarding contraceptive coils prescribed to teenagers]. *Etaplius,* May 11, 2020. https://www.etaplius.lt/valdzios-vyrams-teks-atsakyti-i-klausimus-del-paauglems-skirtos-kontraceptines-spirales.

Bleiker, R. 2015. "Pluralist Methods for Visual Global Politics." *Millennium: Journal of International Studies* 43 (3): 872–890. DOI: 10.1177/0305829815583084.

Budig, M. and P. England. 2001. "The Wage Penalty for Motherhood." *American Sociological Review* 66 (2): 204–225. DOI: 10.2307/2657415.

Burzynska, K. and G. Contreras. 2020. "Gendered Effects of School Closures during the COVID-19 Pandemic." *The Lancet* 395 (10242): 1968. DOI: 10.1016/S0140-6736(20)31377-5.

Catalyst. 2020. "Women in Academia (Quick Take)." *Catalyst*, January 23. https://www.catalyst.org/research/women-in-academia/.

Deakin, H. and K. Wakefield. 2014. "Skype Interviewing: Reflections of Two PhD Researchers." *Qualitative Research* 14 (5): 603–616. DOI: 10.1177/1468794113488126.

Deryugina, T., O. Shurchkov and J. Stearns. 2021. "COVId-19 Disruptions Disproportionately Affect Female Academics." *NBER Working Paper* No. w28360, https://ssrn.com/abstract=3772599.

Gentile, M. 2013. "Meeting the 'Organs': The Tacit Dilemma of Field Research in Authoritarian States." *Area* 45 (4): 426–432. DOI: 10.1111/area.12030.

Hechtman, L., N. Moore, C. Schulkey, A. Miklos, A. Calcagno, R. Aragon and J. Greenburg. 2018. "NIH Funding Longevity by Gender." *Proceedings of the National Academy of Sciences* 115 (31): 7943–7948. DOI: 10.1073/pnas.1800615115.

Howlett, M. 2021. "Looking at the 'Field' through a Zoom Lens: Methodological Reflections on Conducting Online Research during a Global Pandemic." *Qualitative Research* (January): 1–16. DOI: 10.1177/1468794120985691.

Huang, J., A. Gates, R. Sinatra and A. Barabási. 2020. "Historical Comparison of Gender Inequality in Scientific Careers across Countries and Disciplines." *Proceedings of the National Academy of Sciences* 117 (9): 4609–4616. DOI: 10.1073/pnas.1914221117.

Human Rights Watch. 2022. "Turkmenistan." Available at: https://www.hrw.org/europe/central-asia/turkmenistan (accessed on February 15, 2022).

Jowett, A., E. Peel and R. Shaw. 2011. "Online Interviewing in Psychology: Reflections on the Process." *Qualitative Research in Psychology* 8 (4): 354–369. DOI: 10.1080/14780887.2010.500352.

Kamp, M. 2016. "The Soviet Legacy and Women's Rights in Central Asia." *Current History* 115 (783): 270–276. DOI: 10.1525/curh.2016.115.783.270.

Kowal, M., P. Sorokowski, A. Sorokowska, I. Lebuda, A. Groyecka-Bernard, M. Bialek, K. Kowalska, L. Wojtycka, A. Olszewska and M. Karvowski. 2020. "Dread in Academia – How COVID-19 Affects Science and Scientists." *Anthropological Review* 83 (4): 387–394. DOI: 10.2478/anre-2020-0028.

Krause, P., O. Szekely, M. Bloom, C. Fotini, S. Zukerman, C. Lawson et al. 2021. "COVID-19 and Fieldwork: Challenges and Solutions." *PS: Political Science & Politics* 54 (2): 264–269. DOI: 10.1017/S1049096520001754.

Kumenov, A. 2022. "Kazakhstan: Parliament Strips Nazarbayev of Lifetime Sinecures." *eurasianet*, February 2. https://eurasianet.org/kazakhstan-parliament-strips-nazarbayev-of-lifetime-sinecures.

Lalancette, M. and V. Raynauld. 2019. "The Power of Political Image: Justin Trudeau, Instagram and Celebrity Politics." *American Behavioral Scientists* 63 (7): 888–924. DOI: 10.1177/0002764217744838.

Lerchenmüller, C., L. Schmallenbach, A. Jena and M. Lerchenmüller. 2021. "Longitudinal Analyses of Gender Differences in First Authorship Publications Related to COVID-19." *BMJ Open 11* (4): 1–8. DOI: 10.1136/bmjopen-2020-045176.

LRT. 1995. "Panoramos archyvai. Valdžios vyrai rūpinasi šiltu būstu." Trans. [The archives of panorama. The men of the government are taking care of warm housing]. *LRT*, July 14. https://www.lrt.lt/mediateka/irasas/2000185647/panoramos-archyvai-valdzios-vyrai-rupinasi-siltu-bustu.

Malisch, J., B. Harris, S. Sherrer, K. Lewis, S. Shepherd, P. McCarthy, J. Spott, E. Karam, N. Moustaid-Moussa, J. McCrory et al. 2020. "In the Wake of COVID-19, Academia Needs New Solutions to Ensure Gender Equity." *PNAS* 117 (27): 15378–15381. DOI: 10.1073/pnas.2010636117.

Mykolaityte, K. 2020. "Skandalingos medžioklės metu valdžios vyrai nušovė ne tik stumbrę." [During a scandalous hunt men of the government shot not only a female bison]. *lrytas.lt*, February 25. https://www.lrytas.lt/gamta/fauna/2020/02/25/news/skandalingos-medziokles-metu-valdzios-vyrai-nusove-ne-tik-stumbre-13779921.

Nazar, N. 2018. "How Turkmenistan Spies on Its Citizens at Home and Abroad." *openDemocracy*, August 16. https://www.opendemocracy.net/en/odr/how-turkmenistan-spies-on-its-citizens/.

Pannier, B. 2006. "Turkmenistan: Niyazov's Death Leaves Huge Power Vacuum." *Radio Free Europe Radio Liberty*, December 21. https://www.rferl.org/a/1073608.html.

Pateman, C. 1995. *The Disorder of Women: Democracy, Feminism and Political Theory*. Cambridge: Polity.

Pepping, A. and B. Maniam. 2020. "The Motherhood Penalty." *Journal of Business and Behavioral Sciences* 32 (2): 110–125. https://www.proquest.com/docview/2479491839?accountid=8018&pq-origsite=primo.

Polyakova, A. and C. Meserole. 2019. "Exporting Digital Authoritarianism: The Russian and Chinese Models." *Policy Brief.*

Radio Free Europe Radio Liberty. 2019. "German Teach Firm's Turkmen Ties Trigger Surveillance Concerns." *Radio Free Europe Radio Liberty,* February 8. https://www.rferl.org/a/german-tech-firm-s-turkmen-ties-trigger-surveillance-concerns/29759911.html.

Radio Free Europe Radio Liberty. 2021. "VPNs Are Not A-OK: Turkmen Internet Users Forced to Swear on Koran They Won't Use Them." *Radio Free Europe Radio Liberty*, August 10. https://www.rferl.org/a/turkmenistan-vpn-koran-ban/31402718.html.

Radio Free Europe Radio Liberty. 2022. "Serdar Berdymukhammedov Takes the Reins as Turkmenistan's President." *Radio Free Europe Radio Liberty*, March 19. https://www.rferl.org/a/serdar-berdymukhammedov-turkmenistan-president/31760986.html.

Reporters Without Borders. 2019. "2019 World Press Freedom Index." *Reporters Without Borders*, https://rsf.org/en/ranking/2019.

Reuters. 2012. "Belarus Lukashenko: Better a Dictator than Gay." *Reuters*, March. https://www.reuters.com/article/us-belarus-dicator/belaruss-lukashenko-better-a-dictator-than-gay-idUSTRE8230T320120304.

Roberts, S. 2012. "Research in Challenging Environments: The Case of Russia's 'Managed Democracy." *Qualitative Research* 13 (3): 337–351. DOI: 10.1177/1468794112451039.

Sah, L., D. Singh and R. Sah. 2020. "Conducting Qualitative Interviews Using Virtual Communication Tools Amid COVID-19 Pandemic: A Learning Opportunity for Future Research." *JNMA J Nepal Medical Association* 58 (232): 1103–1106. DOI: 10.31729/jnma.5738.

Skriptaite, R. 2021. "Political Lion Skin: What It Means for Belarus to Have a Female Opposition Leader?" *Ballots & Bullets | School of Politics & International Relations, University of Nottingham*, March 8. https://nottspolitics.org/2021/03/08/political-lion-skin-what-it-means-for-belarus-to-have-a-female-opposition-leader/.

Sous, A. 2021. "A Year After Fleeing Belarus, Veranika Tsapkala Still Battling Lukashenka's 'Inhumane Regime.'" *Radio Free Europe Radio Liberty*, August 27. https://www.rferl.org/a/belarus-inhumane-tsapkala-interview/31431465.html.

Sperling, V. 2015. *Sex, Politics, and Putin: Political Legitimacy in Russia*. Oxford: Oxford University Press.

The Guardian. 2019. "Kazakhstan Renames Capital Nur-Sultan." *The Guardian*, March 23. https://www.theguardian.com/world/2019/mar/23/kazakhstan-renames-capital-nur-sultan.

Till, K. 2001. "Returning Home and to the Field." *Geographical Review* 91 (1–2): 46–56. DOI: 10.1111/j.1931-0846.2001.tb00457.x.

Turkmenbashy, S. 1995. *Rukhnama*. Ashgabat: Printing Center of Turkmenistan.

Walker, S. 2015. "A Horse, a Horse … Turkmenistan President Honours Himself with Statue." *The Guardian*, May 25. https://www.theguardian.com/world/2015/may/25/horse-turkmenistan-president-statute-berdymukhamedov.

Wright, K., T. Haastrup and R. Guerrina. 2021. "Equalities in Freefall? Ontological Insecurity and the Long-Term Impact of COVID-19 in the Academy." *Gender Work Organization* 28 (1): 163–167. DOI: 10.1111/gwao.12518.

Afterword

Gaining Access to the Field

Allyson Edwards
Bath Spa University

As the contributions to this volume show the challenges of conducting research are multifaceted and scholars' access to the field varies greatly across the different countries. Some regions, such as the Baltics states, joined the European Union after the collapse of the Soviet Union, facilitating early-career scholars' access to archives and interlocutors on the ground (Ademmer and Delcour 2016). In contrast, political turmoil and authoritarian leadership in Russia, Central Asia, and the Caucasus underpin the difficulties that scholars face when accessing the field.

Running a thoughtful risk assessment is always crucial when conducting high-risk research in the more pressured political environments of the former USSR, as the personal stories in this collection illustrate. Depending on the nature of their project, nationalities, and sexuality, early-career scholars can face visa delays, visa denial, and in some cases even detention and a persona non grata status. The case of Alexander Sodiqov, a PhD candidate at Toronto University who was detained on charges of espionage while interviewing civil society leaders in his "home" country, Tajikistan, demonstrates how dangerous it can be for early-career scholars to conduct research on a politically sensitive topic in the more closed authoritarian post-Soviet region (Amani 2014). Acquiring the correct visa is paramount to carrying out a successful research trip, as the story of Laura Marie Sumner, a PhD student from Nottingham University, who was fined and deported for conducting archival research with the wrong visa in Russia shows (Schreck 2015). Like Sodiqov, Sumner was accused by the local authorities for being a "spy" and was told that her activities constituted a "threat to national security." To avoid similar situations early-career scholars are often advised to affiliate themselves with a local institution that will assist them in the visa application process in providing them with an official invitation letter. Alternatively, the School of Russian and Asian studies (SRAS 2022) can be helpful in providing visa invitations at a low cost.

Early-career researchers' access to the post-Soviet field does not only depend on the high-risk nature of their research project but also on the political stability in the respective country. The historic (and recent) political

DOI: 10.4324/9781003144168-12

tensions between East and West, as well as domestic political unrests, continue to alter researchers' ability to access the region. Over the last few years, we have seen growing instability, especially in the less democratic countries of the region (see the *New Eastern Europe* issue no. 1-2, 2022). For instance, since nationwide, peaceful protests broke out following the reelection of President Alexander Lukashenko in August 2020, Belarusian authorities continue to detain and prosecute scholars, students, journalists, and others for nonviolent political expression (Scholars at Risk 2022). The Kazakhstani government has ceased the visa-free travel regime for almost three months in response to the widespread protests that broke out in the country in January 2022. As a result, many students and scholars had to postpone their field research trips. Until today, traveling without an official invitation letter from a local Kazakhstani university puts a scholar and its research collaborators and participants at risk of police investigation and detention.

Likewise, when Vladimir Putin declared war on Ukraine in February 2022, scholars' access to Russia and Ukraine diminished overnight. Those who are not in Ukraine but who have planned to conduct fieldwork in Ukraine are forced to find alternative ways of conducting their research. In addition, it is important to consider the ethical implications of researching the region at current as Ukrainians experience everyday life and demands in a war-zone. Access to Russia as a fieldwork region is also questionable. Sanctions against Russia have limited flights into and out of the country and have targeted the economic infrastructure of the country. Even if students manage to get to Russia, it will be difficult for them to access resources because of Russia's current economic isolation. Beyond the logistics of living within Russia, increase in xenophobia against the West will also prove problematic to researchers trying to do research in the country. Researchers must make serious ethical considerations when deciding to conduct research in Russia, not only for their own well-being but also for the well-being and safety of their potential interlocutors. There is a high probability that the Russian government will detain non-Russian scholars under spy-pretenses, as a form of political leveraging. It may be a while before nonlocal researchers will be granted access to conduct field research in both Ukraine and Russia.

In addition to the recent political and social turmoil, the COVID-19 pandemic has further impacted scholars' access to the field; almost overnight, teaching, student learning, and research as we knew it moved online. While some contributors in this volume finished their fieldwork prior to the outbreak of the coronavirus in December 2019, those authors who had to adapt and rethink their field research methodology due to the pandemic coped surprisingly well with the unknown situation. For some remote fieldwork presented fewer financial and childcare hurdles than face-to-face interviews. Others noted that pivoting to online interviews made interlocutors open up more during the interview, possibly motivated by boredom during the lockdown and feeling more comfortable by being interviewed in their own space. Switching to remote fieldwork also allowed some of the authors to enter field sites and conduct interviews with people who under normal circumstances, due to the political sensitivity of the study's topic or the researchers' geographical proximity, would not have agreed to the interview.

Nevertheless, remote field research also raises new ethical issues as the stories in this edited volume highlight. For example, not everyone, and in particular scholars in the Global South, enjoys equal access to the Internet because they are missing the necessary digital and media literacy as well as financial means. It is therefore important to note that our authors' experiences might not be shared by scholars who are affiliated to non-elite universities in the West and institutions in the Global South, who do not enjoy the same institutional and financial support. Moreover, for some researchers, remote data collection is not a viable option because their interlocutors lack reliable Internet connectivity or the necessary technology, such as mobile phones and computers, to be interviewed online. Moreover, for scholars who have not been to the field and therefore have not made connections in-person with people of their study community yet, the recruitment of research participants and the organization of online interviews can be even more difficult as the personal account of Skriptaite in this collection shows. Some respondents might be more reluctant to agree to a virtual gathering, due to the risk of online surveillance, hacking, or wiretapping, especially in the more pressured political environment of the post-Soviet region.

Further research on this topic is thus needed, with more emphasis on achieving a wider and broader scope of responses in terms of both the globality and experiences of scholars. However, at this stage, this volume provides an understanding of how some women's access to the post-Soviet field were affected by the pandemic. We hope that some of their reflections on conducting research during the pandemic will help other academics as they move forward with their own research plans in a post-pandemic world.

References

Ademmer, E. and L. Delcour. 2016. "With a Little Help from Russia? The European Union and Visa Liberalization with Post-Soviet States." *Eurasian Geography and Economics* 57 (1): 89–112. DOI: 10.1080/15387216.2016.1178157.

Amani, A. 2014. "A Post-Soviet Dictatorship vs. and Academic Researcher." *openDemocracy*, July 4. https://www.opendemocracy.net/en/postsoviet-dictatorship-vs-academic-researcher/.

New Eastern Europe issue 1-2/2022: "Tug of War? Addressing the Challenge of Instability in the Region." Available at: https://neweasterneurope.eu/2022/02/16/issue-1-2-2022-tug-of-war-addressing-the-challenge-of-instability-in-the-region/ (accessed on March 30, 2022).

Scholars at Risk. 2022. "Protecting Scholars and the Freedom to Think, Question, and Share Ideas." Available at: https://www.scholarsatrisk.org/regions/europe/belarus/ (accessed on March 30, 2022).

Schreck, C. 2015. "Western Scholars Alarmed by Russian Deportations, Fines." *RadioFreeEurope*, March 31. https://www.rferl.org/a/russia-western-scholars-alarmed-deportations/26929921.html.

SRAS. 2022. "School of Russian and Asian Studies. Study, Research, and Custom Programs Abroad." Available at: https://sras.org/ (accessed on March 30, 2022).

Index